CONCRETE BRIDGE ENGINEERING

Performance and Advances

CONCRETE BRIDGE ENGINEERING
Performance and Advances

Edited by

R. J. COPE

Plymouth Polytechnic, UK

ELSEVIER APPLIED SCIENCE
LONDON and NEW YORK

ELSEVIER APPLIED SCIENCE PUBLISHERS LTD
Crown House, Linton Road, Barking, Essex IG11 8JU, England

Sole Distributor in the USA and Canada
ELSEVIER SCIENCE PUBLISHING CO., INC.
52 Vanderbilt Avenue, New York, NY 10017, USA

WITH 23 TABLES AND 158 ILLUSTRATIONS

© ELSEVIER APPLIED SCIENCE PUBLISHERS LTD 1987

British Library Cataloguing in Publication Data

Concrete bridge engineering: performance and
advances.
1. Bridges, Concrete
I. Cope, R. J.
624'.2   TG335

Library of Congress Cataloging in Publication Data

Concrete bridge engineering.

Includes bibliographies and index.
1. Bridges, Concrete—Design and construction.
2. Bridges, Concrete—Maintenance and repair.
I. Cope, R. J. (Robert J.)
TG335.C57   1987   624'.2   87-13672

ISBN 1-85166-110-7

Printed in Great Britain by Galliard (Printers) Ltd, Great Yarmouth

To John and James

Preface

Concrete bridges have played a vital role in enabling the growth of the communication facilities demanded in the second half of the twentieth century to be achieved. It is difficult to imagine how the needs for greater speed and volume of goods transported could have been met without the rapid developments in design and construction technologies that took place between the 1950s and the 1970s. So fast were those developments that some risks had, inevitably, to be taken. Partly as a result of the need to provide a large number of structures in a short time, and partly due to lack of adequate research and development funding over that period, the growing technologies of the 1980s are as much concerned with assessment, repair and maintenance of existing structures, as they are with the development of new concretes and with improved methods of design and construction for new structures.

This book describes the latest methods for assessing and repairing concrete bridges, and provides information on developments in materials and design methods for future bridges. Bridge decks, joints and substructures are all considered. Each invited chapter is written by an eminent international authority. In compiling and editing the text, the writer came to appreciate that so vast is the range of technologies now used by bridge engineers, that there are likely to be only a limited number of single-authored technical texts on bridge engineering in the future. The information provided is distilled for practising bridge engineers, researchers and postgraduate students from both long-running research programmes and the experiences of very senior engineers. The text provides a state-of-the-art report on major aspects of developments in design and materials and in methods for assessment and repair of concrete bridge structures.

R. J. COPE

Contents

List of Contributors

H. Aizlewood

  158 Baginton Road, Coventry CV3 6FT, UK

B. deV. Batchelor

  Department of Civil Engineering, Queen's University, Kingston, Ontario, Canada K7L 3N6

P. Bennison

  Liquid Plastics Ltd, PO Box 7, London Road, Preston PR1 4AJ, UK

J. H. Bungey

  Department of Civil Engineering, University of Liverpool, PO Box 147, Liverpool L69 3BX, UK

J. L. Clarke

  Cement and Concrete Association, Wexham Springs, Slough SL3 6PL, UK

R. J. Cope

  Department of Civil Engineering, Plymouth Polytechnic, Plymouth PL1 2DE, UK

M. J. N. Priestley

Department of Applied Mechanics and Engineering Sciences, University of California, La Jolla, San Diego, California 92093, USA

H. P. J. Taylor

Dow-Mac Concrete Ltd, Tallington, Stamford, Lincolnshire PE9 4RL, UK

1

Planning Site Investigations

J. H. BUNGEY

Department of Civil Engineering,
University of Liverpool, UK

1.1 INTRODUCTION

On-site investigation of concrete bridges may be necessary for a wide range of reasons which will generally be associated with assessment of either specification compliance, maintenance requirements or structural adequacy. Establishing the precise aims of inspection and testing is an essential prerequisite to all test programme planning including fundamental aspects such as selection of methods, location of test points and interpretation of results. On-site inspection and testing, other than superficial visual inspection, is seldom cheap as complex access arrangements are frequently necessary and procedures may be time-consuming. Furthermore, individual observations or test results are often inconclusive, and back-up testing coupled with considerable engineering judgement and experience are then required.

The importance of careful planning cannot be overemphasised, if the maximum amount of worthwhile information is to be obtained at minimum cost. Ideally, a programme should be planned to evolve sequentially, as illustrated in Fig. 1.1, in the light of results obtained in its earlier stages and using only those stages necessary to achieve conclusions with an adequate degree of confidence. Although it is recognised that this may sometimes pose financial difficulties in that the cost will not be clearly defined at the outset, there is little doubt that this will generally provide the most cost-effective approach. It should also be noted that objectives, as originally defined, may change in the course of an investigation, possibly as the result of litigation, and both planning and documentation should

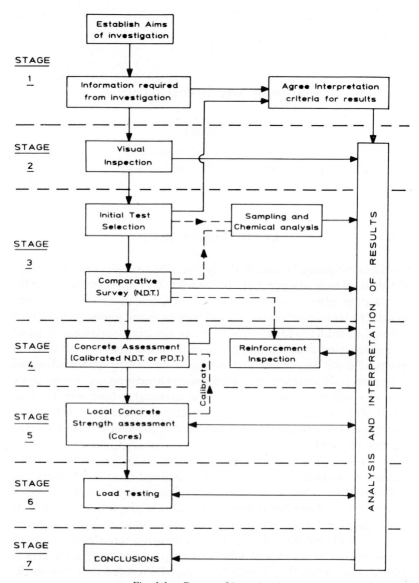

Fig. 1.1. Stages of investigation.

attempt to bear this eventuality in mind. A further component of a successful site investigation is that full agreement is reached between all parties involved concerning the interpretation of results before any testing commences. This is essential if future disputes about the significance of results are to be avoided.

Relevant testing methods and equipment are described in Chapter 2, to which reference should be made for further detailed sources of information about individual techniques.

1.2 TYPES OF INVESTIGATION

1.2.1 Specification Compliance
Determination of compliance of concrete bridges with their specification is primarily undertaken during, or soon after, construction, and only exceptionally will such testing be required at later stages in the life of a structure.

1.2.1.1 Routine Testing at Time of Construction
This will consist, in part, of quality control testing of materials, which usually utilises representative samples of the materials to be used. In the case of concrete, however, the 'standard' 28-day compressive strength specimen in the form of a cube or cylinder cannot take account of the levels of compaction or curing received by the concrete in the structure. Furthermore, the limited number of samples for this purpose, or for 'fresh' properties such as workability or air content, which are normally associated with an individual pour cannot ever truly reflect variations of the material actually used, even if samples are taken according to a very well controlled plan. There is a growing awareness in some parts of the world (notably the USA and Scandinavia) of the benefits of physical control testing of the *in-situ* concrete, in addition to standard specimens. Appropriate non-destructive methods for this type of investigation include pull-out, penetration resistance, break-off, and ultrasonic pulse velocity surveys.

The other principal aspects of 'investigation' at early ages of a structure are essentially visual, concentrating on line, level and surface appearance of the concrete following formwork removal. Unfortunately this gives little indication of the likely performance of the structure since interior defects, including incorrect positioning or displacement of reinforcement, cannot be seen. In view of the numerous reported cases of inadequate thickness of concrete cover and the subsequent effects of this upon long term durability, there is a very strong case for measurement of cover as a matter of routine following removal of formwork to ensure that specified thicknesses have been provided. Such testing is quick and straightforward, adding little to the cost of a project and yet potentially offering significant long term advantages to a client.

Other tests which are available include analysis of the fresh concrete to determine cement content, and surface absorption measurements; however, neither of these approaches is currently applied to bridge construction in the United Kingdom as a matter of routine. It is of interest to note, however, that specifications in Denmark may call for microscopic thin-section analysis of samples obtained from cores taken from newly cast concrete in order to confirm cement type and content.

1.2.1.2 Non-routine Testing Following Construction

Such testing will normally arise from failure of 'standard' compression specimens to achieve the required strength. Testing will thus be aimed at confirming or disproving that suspect concrete is substandard, and if necessary determining the extent of such material. It is in such situations that the use of comparative non-destructive testing followed as necessary by more expensive and disruptive techniques, as outlined in Fig. 1.1, is of greatest value.

Tests of concrete strength, of which core tests are the most reliable, are able to assess the actual *in-situ* value at a given location to within 95% confidence limits of $\pm 12/\sqrt{n}\%$ where n samples are tested.[1] Whilst this may appear to be very useful, uncertainties concerning the relationship between actual *in-situ* strength and that of 'standard' specimens (sometimes referred to as 'potential' strength) complicate interpretation. Marginal cases are thus difficult to prove or disprove with any degree of certainty.

It is essential that such testing is designed to obtain results relating to 'average' conditions within the suspect part of the structure whilst recognising the likely variations and level of *in-situ* properties appropriate to the particular member type, as discussed later. Locations of tests for this purpose will thus not necessarily correspond with those which may be required to determine structural adequacy, and it is important that this feature of test planning is recognised.

1.2.1.3 Testing at Later Stages

Testing of this type will usually result from abnormal deterioration being observed in the course of routine inspections, and will often be associated with litigation. Chemical analysis or microscopic thin-section analysis of appropriate samples taken from the *in-situ* concrete will frequently be used to attempt to determine mix features such as cement type and content, chloride or sulphate levels, entrained air content, or excessive admixture dosage. In embarking upon a test programme for this purpose it is again important to recognise the limits upon accuracy within which the desired property can be assessed.[2] Normal material variability will mean that extensive sampling may be required and it is usually only gross deficiencies that may be identified with a sufficient degree of confidence to prove non-compliance with specification limits.

Tests of concrete *in-situ* strength will be subject to the problems of interpretation indicated above, further aggravated by age and moisture effects, and it will again usually only be possible to confidently demonstrate non-compliance with strength specifications in cases of major shortfall.

The other major testing approach within this category involves measurement of the thickness of concrete cover to embedded reinforcement or prestressing ducts. This can be quickly achieved non-destructively by the use of electromagnetic cover measuring devices which are generally accurate to within about ± 5 mm when properly calibrated for the particular concrete and steel characteristics. In this case, however, it is a simple matter to obtain conclusive evidence about the adequacy of cover values by drilling or breaking out at suspect areas identified by the non-destructive tests. This particular equipment may also sometimes be useful to confirm the number and spacing of bars present in cases where doubts exist.

1.2.2 Routine Inspection

The need for necessary remedial work to be undertaken in good time to prevent more serious and costly structural deterioration is particularly acute in the case of bridges. This is due to a combination of the exposure conditions to which most bridges are subjected, leading to excalating deterioration rates, and the severe social and commercial costs and consequences generally resulting from closure or application of load restrictions. Economic maintenance is thus heavily dependent upon regular and thorough inspection and monitoring to detect deterioration at an early stage. It is important that such inspections are programmed and executed in a systematic manner, and requirements relating to motorway and trunk road bridges in the United Kingdom are set out in the Department of Transport's Standard BD22/84.[3] This document identifies four basic types of inspection.

1.2.2.1 Superficial Inspection

This should be an ongoing process, with staff encouraged to report any obvious deficiencies which may lead to more serious long term problems or loss of safety to the bridge users. Inspections of this type would include features such as impact or flood damage and expansion joint deterioration. Reports of such problems may also originate from bridge users, and it is essential that procedures are established for reports to be followed up and properly documented.

1.2.2.2 General Inspection

Representative parts of the structure should be inspected visually every one or two years. This will not generally require any specialised access equipment, although binoculars or telescopes may sometimes be useful aids. Features to be observed will include deflections and distortions, cracking, surface deterioration and leakage. The results of these inspections should be made on standardised forms to simplify data analysis and storage of inspection records, and this requires classification of both the extent and severity of observed defects and an indication of their likely cause. Inspectors are also required to give a repair priority assessment and a cost estimate of each item of work identified.

1.2.2.3 Principal Inspections

These are required every six to ten years and represent a major undertaking involving access equipment with possible lane closures and railway track possessions. This inspection requires a close examination of all inspectable parts of the structure and will be predominantly visual, aided by devices such as crack-width microscopes and fibre-optic equipment. Where defects are found, however, the subsequent detailed investigation to assess their extent and cause may involve a wide range of equipment. This may include assessment of features such as cover, reinforcement corrosion, internal concrete integrity and defects, and material properties. The results may possibly lead to an assessment of structural adequacy as described below. Long term monitoring of structures for crack development, settlement and subsidence may also follow from such inspections.

Results of principal inspections are reported in a similar manner to those of general inspections but in the case of initial inspections, or the first principal inspection following radical structural changes, the report should be supplemented by standardised details of construction of the structure to ensure proper documentation for future use.

1.2.2.4 Special Inspections
Special inspections are necessary for bridges which are identified as being
subject to particular risk. In the case of concrete structures this includes
those which

    (a)   exhibit a specific condition causing concern,
    (b)   are subject to load restriction,
    (c)   are at risk from subsidence in mining areas, or
    (d)   are required to carry an abnormal load which exceeds that already
        documented or is likely to induce critical stresses.

Other cases requiring special inspection include those where excessive
settlement is observed, following major fires under structures, and
following possible flood damage to foundations.

The nature of special inspections will vary widely according to their
purpose. They may make use of a wide range of the testing techniques
described in Chapter 2 and may lead to an assessment of structural
adequacy.

Routine inspection reports are the principal source of information from
which maintenance work is planned, and it is thus essential that meticulous
documentation is maintained in an accessible format, both to assist
maintenance programming and to ensure that the performance of the
structure can be efficiently monitored thoughout its life. The significance of
this task is highlighted by the fact that there are over 150 000 highway
bridges alone in the United Kingdom.

1.2.3 Structural Adequacy
Assessment of structural adequacy is likely to follow from failure to comply
with the requirements of specifications, observation of defects or damage
during routine inspections, or the need for a structure to carry abnormal or
upgraded loadings. The planning of the programme of testing will follow
the general procedures described in the following section of this chapter,
but will depend to a large extent upon both the nature of the problem and
the level of existing documentation. Where little documentation is available
the testing may include dimensional measurements, determination of
structural actions, reinforcement identification and location, concrete
materials identification and properties assessment, and may culminate, in
extreme cases, with load testing.

Theoretical load capacity assessment is outside the scope of this chapter,
but such calculations will require appropriate values of materials
properties. The United Kingdom Department of Transport Departmental

Standard BD21/84[4] gives advice relating to the strength assessment of certain types of highway structures under normal loadings, and assessment of reinforced concrete slab bridges is considered in greater detail in Chapter 3.

Assessment of reinforcement characteristics will usually be relatively straightforward, involving, if necessary, tensile testing of representative samples removed from the structure. Concrete properties are, however, much less easy to determine. The properties of most interest are usually the elastic modulus and the strength.

Dynamic modulus. This may often be determined with sufficient accuracy from *in-situ* ultrasonic pulse velocity measurements or, if this is not possible, from laboratory tests on cores removed from the structure.

Strength. As discussed previously, this may be achieved most reliably by the use of cores, although a range of 'partially destructive' *in-situ* surface zone tests are also available.

The elastic modulus and the strength of concrete will vary, both in a random manner and according to member type, as discussed in detail below. Difficulties thus arise concerning the locations of tests to obtain values appropriate to the particular calculations in hand. The variations in the elastic modulus are unlikely to be as marked as those of strength; thus an 'average' value is likely to be adequate for calculations of overall structural behaviour, although 'extreme value' estimates may be worthwhile in critical cases. Strength calculations for individual structural components should, however, be based on values relating to regions which are critically stressed and to those that are likely to exhibit the lowest concrete strength, and test points must be located accordingly. Nondestructive or partially destructive testing may be useful in confirming locations of lowest concrete properties. Where calculations involve a value of concrete strength obtained from *in-situ* measurement, the questions of relationships between actual *in-situ* values and those of standard specimens (upon which calculations are normally based) and appropriate factors of safety arise. These problems are discussed fully in Section 1.4.4.

An important feature of structural adequacy assessments is that they relate only to one point in time. If deterioration has occurred, it is essential to establish the cause, and whether it has ceased or is likely to continue. This is particularly relevant to load tests, which provide an excellent demonstration of ability to carry a particular loading but give little or no indication of future reserve of strength.

1.3 TEST PROGRAMME PREPARATION AND EXECUTION

The need for a programme to be subdivided into stages and to be able to evolve has already been identified and illustrated in Fig. 1.1, whilst details of particular test methods are given in Chapter 2. There are, however, a number of other important general features of planning which must be considered, and these are summarised in Fig. 1.2.

The Institution of Structural Engineers document *Appraisal of Existing Structures*,[5] although principally concerned with buildings, contains much useful information concerning procedures, appraisal processes and methods, as well as determination of testing requirements. Guidance is also offered on sources of information, reporting, and identification of defects with their possible causes, and appropriate suggestions for investigation.

1.3.1 Documentation

Having established the initial objectives of the investigation, all likely sources of relevant documentation should be identified as quickly as possible. Available site records, drawings, contract documents, materials reports, environmental records and inspection reports should be studied to provide as much background data as possible before any site visit, to permit maximum benefit from the visit. In practice, however, full documentation will seldom be available and it will often be necessary to start an investigation with incomplete background knowledge.

1.3.2 Preliminary Site Visit

This is important, not only to obtain a feel for the problem before detailed planning but also to permit an initial assessment of practical factors influencing the choice of test methods as well as identification of safety and access requirements.

1.3.3 Access and Safety

Provision of adequate access for both inspection and testing is frequently one of the most difficult and expensive aspects of an investigation. Access equipment ranges from simple ladders or scaffolding to specialised rail-hung travelling gantries or cradles. Mobile aerial platforms (with the platform above or below vehicle level) involve a variety of telescopic or articulated arm arrangements. Boats may be necessary in some cases, possibly taking advantage of extremes of tidal conditions to gain close access to various parts of the structure, whilst in other extreme cases

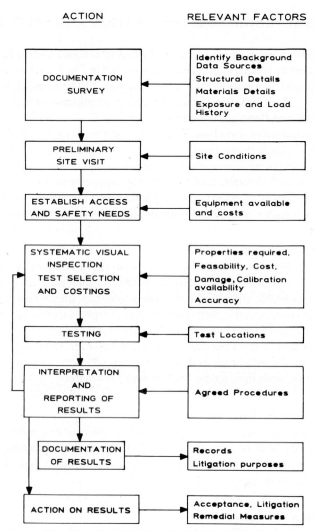

Fig. 1.2.   Detailed planning of an investigation.

specialist mountaineering techniques such as abseiling may be advantageous. Underwater surveys will generally be carried out by specialist diving teams, possibly using techniques such as TV scanning, and it is important to identify the extent of site back-up base facilities required in these situations. The choice of appropriate access arrangements will depend upon several features relating to the site as well as the number of personnel and types of test equipment to be used. Useful guidance has been provided by George[6] which may assist such decisions as well as giving information about worthwhile 'personal' equipment for those performing the inspection.

Safety, both of the inspection personnel and the general public, is clearly of utmost importance. Appropriate traffic signs may be necessary for work on highway structures, whilst the safety provisions of the various statutory requirements such as the Health and Safety at Work Act and the Factories Act must always be observed. Work on structures associated with highways, railways or waterways will also be subject to the safety requirements of the relevant authorities. Safety harnesses or life jackets may be necessary in some situations, whilst for work over deep water safety boats may be required. Although extensive safety precautions may be necessary for large scale radiography,[7] testing itself will not generally cause public danger. It is important, however, that care is taken to ensure that test personnel use appropriate protective clothing and equipment, and that proper care is taken over the use of electrical equipment on site. Where load testing is undertaken, precautions will be necessary to guard against total collapse, and these precautions may be both extensive and expensive. All possible structural responses to the test load should be examined and safety and access requirements determined accordingly.

Legal liability, indemnity and insurance aspects of the proposed programme of testing are a further important feature which should be discussed between all parties, with responsibilities established at an early stage.

1.3.4 Test Method Selection

Detailed selection of test methods will be based on a knowledge of the established aims of the investigation, coupled with a knowledge of access and practical restrictions obtained from the preliminary site visit. A detailed systematic visual inspection will usually be necessary to establish the precise locations and extent of testing and to help identify causes of deterioration, although this may not be possible until the time of testing due to the need for access provision.

Important considerations in the selection of methods will include:

The availability and reliability of calibrations, which may be required to relate measured values to the required properties. In some cases it may be necessary to break out concrete for visual examination or to cut cores for crushing to achieve these calibrations.

The effect of damage, which will relate to both the surface appearance of the test member and the likelihood of structural damage resulting from the testing of small-section members.

Practical limitations. Important features will include the member size and type, surface condition, depth of test zone, location of reinforcement and access to test points. Other factors may also include ease of transport of equipment, effect of environment on test methods, and safety of test personnel and the general public during testing.

The accuracy required, which will influence not only the choice of test method but also the number of test points to provide meaningful results.

Economics. The value of the work under examination and the cost of delays must be carefully related to the likely cost of a particular test programme. The available budget may also be a constraint influencing the choice of methods and the extent of testing possible.

The benefits of organising testing in a sequence, such as that suggested in Fig. 1.1, in which the tests involving least cost and damage are used initially and are followed by other methods as necessary, has already been emphasised. In particular, preliminary comparative surveys using non-destructive methods of relatively low cost are often worthwhile where the investigation is concerned with material properties or conditions. In this way, greatest benefit can be obtained from a limited number of higher precision but more expensive or disruptive tests. Cores are especially valuable as samples for a wide range of laboratory tests relating to material characteristics and durability.[8]

The results of each stage of testing should be analysed in the light of the agreed requirements before proceeding further. Combinations of test methods may be particularly useful to confirm observed patterns of results or to increase the reliability of estimated properties.[9]

It is important to recognise that some methods are particularly sensitive to variations of testing procedure, whilst in many instances it is only possible to obtain approximate estimates of the required properties by comparative means. Skill and care by the operator will always be necessary, and reliable trained and experienced staff must always be used despite the apparent simplicity of some methods.

1.3.5 Testing
The two most fundamental aspects of a test programme, which should always be carefully executed in accordance with current established

Standards or accepted practice, are the location and the number of test points which are to form the basis of the sampling plan.

1.3.5.1 Test Positions

These will depend entirely upon the purpose of the investigation as discussed previously. Whilst testing will often be concentrated upon parts of the structure which are for some reason suspect or showing signs of deterioration, it is important that tests for specification compliance attempt to obtain representative average results for the relevant members. Tests for structural adequacy will concentrate upon areas which are critical from the point of view of high stress and lowest strength capacity. Attention in these cases will thus often be concentrated on the upper zones of members unless other regions are particularly suspect. If aspects of durability are involved in the investigation, care should be taken to allow for variations in environmental exposure conditions between different parts of the structure or member under test. Test positions must also take into account the possible effects of reinforcement upon results (and if necessary individual test points be arranged to avoid steel) as well as any physical restrictions imposed by a particular method in use.

The number of load tests that can be undertaken on a structure will be limited and these should be concentrated on critical or suspect areas. Visual inspection and non-destructive tests may be valuable in locating such regions. Where individual members are to be tested destructively to provide a calibration for non-destructive methods, they should preferably be selected to cover as wide a range of concrete quality as possible.

Test positions must always be clearly measured and identified to permit proper interpretation and documentation of results.

1.3.5.2 Number of Tests

Establishing the appropriate number of tests is inevitably a compromise between the accuracy required and the effort, time, cost and damage involved. Mathematical procedures may be used to evaluate the number of individual tests needed to achieve a specified level of accuracy, taking account of testing and material variability,[10,11] but, in practice, it will usually be necessary to settle for fewer readings than this theoretical ideal, coupled with lower accuracy.

Test repeatability varies widely according to method and will determine the number of individual tests required at a location to obtain a reliable average value for that location (e.g., 1 UPV minimum value measurement, 3 Windsor probes, 10 rebound hammer values). The accuracies of prediction of correlated properties based on such values are discussed below, and in greater detail in Chapter 2 and elsewhere.[2] Random material variability will influence the number of test locations which must be examined to assess concrete which is expected to be uniform in *in-situ* quality and exposure history characteristics. *In-situ* patterns of concrete quality according to member type will also influence the number of locations to be tested if the quality of an entire member is to be assessed comprehensively.

For comparative purposes of concrete quality assessment, the non-destructive methods are the most efficient since their speed permits a large number of locations to be easily tested. For a survey of concrete within an individual member at least 40 locations are suggested, spread on a regular grid over the member; whilst for comparison of similar members a smaller number of points on each member, but at comparable positions, should be examined. Where it is necessary to resort to other methods, such as partially destructive tests, practicalities are more likely to restrict the number of locations examined, and the survey may be less comprehensive.

The number of 'standard' cores necessary to achieve a given accuracy of *in-situ* concrete strength has already been discussed, and where cores are being used to provide a direct indication of strength as a basis of calibration for other methods, it is important that enough cores are taken to provide an adequate overall accuracy. It is also essential to recognise that the results will relate only to the particular locations tested and these should, therefore, be selected to provide maximum benefit. The added complications associated with determination of specification compliance are discussed later, and a minimum of four cores from a suspect batch of concrete is recommended.[1] Where 'small' cores are used, a greater number will be required to give a comparable accuracy, due to greater test variability.[2]

Surveys relating to reinforcement corrosion, such as cover measurement or half-cell potential or pulse-echo measurements, will normally be undertaken on an initial grid related to reinforcement spacings and member size. The grid spacing may then be reduced for closer examination of suspect areas which are identified. If resistivity measurements are to be made, these will also be concentrated in these suspect areas which will additionally be prime locations for carbonation and chloride analysis tests.

The number of chemical tests required will also be largely determined by the need to obtain representative values, bearing in mind likely material variability, member size and the extent to which results may be realistically extrapolated. This problem is discussed more fully in Chapter 2, and although the necessary number of tests or samples will be a matter of engineering judgement, the criteria discussed above for strength testing should provide a useful basis.

1.3.6 Interpretation of Results

Detailed aspects of interpretation of results are discussed later. It is particularly important, however, that preliminary analysis and interpretation are continuing processes throughout the site stages of the investigation, as shown in Fig. 1.1. This will permit the most efficient use of resources on site, and lead to maximisation of the value of the information obtained during a period which is often restricted by access provisions. To this end, it is essential that an appropriately experienced engineer is available to assess results on an ongoing basis and who has the authority to modify the programme according to the specific requirements of the particular bridge under examination.

1.3.7 Documentation of Results

Procedures for reporting the results of routine inspections to provide a data base for future reference have already been outlined. The need for comprehensive and detailed recording and reporting for investigations of other types is of equal significance, no matter how small or straightforward the problem may at first appear to be. In the event of subsequent dispute or litigation, the smallest detail of procedure may be crucial and records should always be kept with this in mind. Comprehensive photographs of the structure and features under examination are often of particular value for future reference, whilst the technique of crack-mapping to monitor the development, and identify the causes, of deterioration is discussed in Chapter 2.

Particular features of the site investigation report will generally include:

(a) the date, time and place of test, with details of environmental conditions;

(b) a description of the structure and its history of load and environmental exposure, and details of any modifications or repairs;

(c) details of the test procedures and equipment used including reference to relevant British or other Standards, drawing attention to features which do not comply with their recommendations;

(d) locations of test points, illustrated by dimensioned sketches;

(e) details of concrete and conditions of test, with particular attention to features known to influence the results of tests used;

(f)   mean, range, standard deviation and coefficient of variation of measured values as appropriate;

(g)   test results expressed in terms of a correlated property, including details of the correlation source.

Other specific requirements of reports for particular test methods are given in the appropriate Standards.

1.4   INTERPRETATION

The importance of interpretation as a process which must continue throughout any investigation has been emphasised. This will range from qualitative judgements concerning features observed during visual surveys, to detailed analysis and statistical evaluation of numerical test results with subsequent quantitative assessment of physical properties leading to the formulation of conclusions.

Assessment of the results of visual inspections will rely heavily upon the skill and subjective judgement of the engineer performing the inspection. Numerical classification is generally limited to placing various observed features within 'zones' according to condition as used, for example, by the Danish Ministry of Transport[8] for crack classification. Roper *et al.*[12] have however recently outlined an approach using cross-cause flow charts and statistical techniques applied to such classifications. It is claimed that this method can be developed to provide a systematic quantitative measure of durability.

1.4.1   Initial Computation of Numerical Test Results

The amount of computation required to provide the appropriate parameter at a test location will vary according to the test method but will follow well-defined procedures. For example, cores must be corrected for length, orientation and reinforcement to yield an equivalent cube strength. Pulse velocities must be calculated making due allowance for reinforcement, whilst pull-out, penetration resistance, surface hardness and similar tests must be averaged to give a mean value. Attempts should not be made at this stage to invoke correlations with a property other than that measured directly. Electrical, chemical or similar tests will be evaluated to yield the appropriate parameter, and load tests will usually be summarised in the form of load/deflection curves with moments evaluated for critical conditions, and with creep and recovery indicated.

1.4.2 Conversion of Test Results to Give a more Useful Parameter

This process will primarily be associated with the use of correlations between measured values and concrete strength, since this is the material property which is most commonly required by engineers concerned with aspects of specification compliance or structural adequacy. Any calibration must be relevant to the material in use, and in some circumstances it may be necessary to develop such relationships specially.

Particular attention must be paid to the differences between laboratory conditions (for which calibration curves will normally be produced) and site conditions. Differences in maturity and moisture conditions are especially relevant in this respect. Concrete quality will vary throughout members and may not necessarily be identical in composition or condition to that of laboratory specimens. Also, the tests may not be so easy to perform or control on site due to adverse weather conditions, difficulties of access or lack of experience of operatives. Calibration of non-destructive and partially destructive strength tests by means of cores from the *in-situ* concrete may often be possible and will reduce some of these differences.

The accuracy of *in-situ* strength prediction will depend both upon the variability of the test method itself and the reliability of the correlations, but even in ideal circumstances with a specific calibration for the mix in use, it is unlikely to be better than the values given in Table 1.1.

Accuracy of strength estimation may sometimes be improved by mathematical combination of results of two separate types of non-destructive[13] or partially destructive[14] tests, each with their appropriate

TABLE 1.1
Maximum Likely Accuracies of *in-situ* Concrete Strength Prediction

Method	Max. likely strength accuracy (95% confidence limits)
Cores (4): 'Large' (> 100 mm)	± 6%
'Small'	± 18%
Pull-out (6)	± 10%
Pull-off (3)	± 15%
Break-off (5)	± 20%
Windsor probe (3)	± 20%
Ultrasonic pulse velocity	± 20%
Rebound hammer (10)	± 25%
Internal fracture (6): Direct pull	± 20%
Torquemeter	± 28%

strength correlations, although this approach tends not to be used to any great extent in the UK at present.

1.4.3 Examination of Variability of Test Results

Whenever more than one test is carried out, an examination of the variability of results can provide valuable information. Even where few results are available, these can provide an indication of the uniformity of the construction and hence the significance of the results. In cases where more numerous results are available, as in non-destructive surveys, a study of variability with the aid of contour plots or histograms can be used to define areas of differing quality.

Under normal circumstances contours may be expected to follow well defined patterns, and any departures from this pattern, or that anticipated for the particular member type, will indicate areas for concern. A typical contour pattern for a section of well constructed reinforced concrete beam is illustrated in Fig. 1.3.

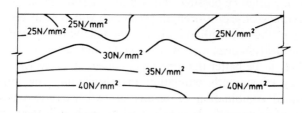

Fig. 1.3.   Typical beam in-situ strength contours.

Normal variability of the supplied material may be expected to be distributed randomly, but compaction and curing effects will be influenced by characteristics of the member under construction. This will tend to lead to strength gradients across the member depth, with general trends as indicated in Fig. 1.4.[2] The basic concept of reduced strength in uppermost zones of members is recognised by BS 6089.[15] There may also be further strength differences of up to 10% between surface zones and the interior of members, resulting from curing effects.

Histograms provide another useful graphical technique for assessing material and construction uniformity, with the spread reflecting the member type and the distribution of test locations as well as construction features.

Evaluation of coefficients of variations of test results is also recommended, since these can be coupled with a knowledge of the variability

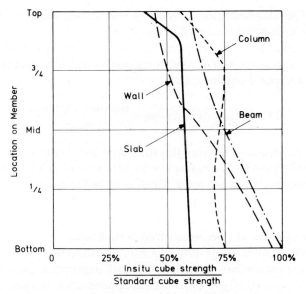

Fig. 1.4. Typical within-member in-situ strength patterns.

associated with the test method to provide a measure of the construction standards and control used. This is illustrated by some typical examples in Table 1.2 and discussed more fully by the author elsewhere.[2]

The values in Table 1.2 relate to a single unit made on site from a number of batches. Results from a single batch would be expected to be lower, whereas if several member types are involved, values may be expected to be

TABLE 1.2
Typical Coefficients of Variation of Test Results on an
Individual Member for Good Construction Quality

Method	Coefficient of variation
Cores: 'Large' (≥ 100 mm)	10%
'Small'	15%
Break-off	10%
Pull-out	7%
Windsor probe	4%
Rebound hammer	4%
Ultrasonic pulse velocity	2.5%
Internal fracture	20%

higher. The tabulated values offer only a very general guide, but should be sufficient to indicate abnormal circumstances.

The coefficient of variation of concrete strength is not likely to be constant for a given level of control because it is calculated on the basis of the mean strength obtained; i.e.

$$\text{Coefficient of variation} = \frac{\text{standard deviation}}{\text{mean}}$$

General relationships between this parameter and construction standards are thus not possible. Values as low as 10% may be achieved for typical bridge concrete[16] but are more likely to be in the range of 15–20% for average construction quality.

1.4.4 Formulation of Conclusions from Test Results

The results of some forms of testing may be used directly as the basis of engineering judgements, in conjunction with appropriate pre-established criteria of acceptability or specification requirements. In these cases, the principal considerations will include the likely accuracy of results and the extent of their general relevance to the body of concrete under investigation. It is not uncommon for marginal results to be inconclusive, in view of the limited accuracy that is possible with many testing approaches and restrictions upon the number of test results that can be obtained. Sometimes the acceptance criteria themselves may also involve an element of uncertainty, for example where durability is concerned. Under these circumstances it may well be necessary to resort to further investigation using a complementary testing approach in an attempt to confirm observed patterns of behaviour and increase confidence.[9]

The use of *in-situ* strength test results poses particular problems in that specifications and calculations are almost always based upon characteristic strengths of standard specimens cured and tested at 28 days at 20°C under moist conditions. Differences of compaction, moisture condition, curing and age will inevitably mean that *in-situ* concrete strengths are different from those achieved by standard specimens of the same concrete. In design, this is often allowed for by the use of a generalised partial factor of safety on concrete strength, but in practice, the differences vary according to member type and location within the member.

Figure 1.4 illustrates typical differences, but it must be emphasised that these may be subject to considerable variations in individual cases. In particular, the differentiation between beams and slabs is ill-defined and will depend upon thickness and construction techniques. Figure 1.4, however, demonstrates the difficulties and uncertainties involved in attempting to predict a corresponding standard cube strength (or 'potential' strength) from *in-situ* test results. Consequently, there is likely to be a considerable 'unproven' zone when considering strength specification compliance, even when test locations have been selected to give representative 'average' results for the member. The number of *in-situ* test results will seldom be sufficient to permit proper statistical analysis to determine the appropriate 'characteristic' value. Hence, it is better to compare mean *in-situ* strength estimates with the expected mean 'standard' test specimen result.[2] This requires an estimate to be made of the likely standard deviation of standard specimens, unless the value of target mean strength for the mix is known.

In-situ strength values measured at a critical location for calculations of structural adequacy are best used in the form of a mean value for the location, with a factor of safety applied to this to allow for test variability, lack of concrete homogeneity and future deterioration. The accuracy of strength prediction will vary according to the method used, but a factor of safety of not less than 1·2 is recommended by BS 6089[15] for general use. If there is particular doubt about the reliability of the test results, or if the concrete tested is not from the critical location considered, then it may be necessary for the engineer to adopt a higher value for the factor of safety. Alternatively, other features, including moisture condition and age, may possibly be used to justify a modified value for the factor of safety, whilst the *in-situ* stress state and rate of loading may also be taken into account in critical circumstances.

Further detailed guidance relating to calculations to assess specification compliance and structural capacity from *in-situ* strength results has been provided by the author elsewhere.[2]

1.5   GENERAL OBSERVATIONS

The need for regular and systematic inspection of concrete structures is recognised internationally, and a recent FIP Report[17] recommends general classifications and procedures complementing those described above. Because of their critical nature and exposed environment, bridges are particularly susceptible to deterioration and their inspection and testing is thus of especial importance. However, access to all parts of the structure, including critical zones such as bearings and joints, frequently presents major difficulties and expense. Whilst advances have been made in the development of equipment to assist these operations, it is essential that greater attention is paid to this requirement at the design stage of future bridges. The other key feature of site inspections and investigations is the preparation and maintenance of thorough records and reports. Computer data storage and retrieval systems may offer scope for developments in this area, and may lead to the incorporation of investigation reports into computer based maintenance management systems.[18]

The desirability of systematic but flexible planning of test programmes which go beyond routine inspection has been emphasised, but acceptance of this approach by all parties often presents difficulties, as does agreement upon interpretation of results. The latter situation may perhaps be expected to improve with greater education emphasis upon the *in-situ* characteristics of concrete. Whilst experienced testing engineers recognise the potential long term economic benefits if investigations are allowed to develop in stages, commissioning engineers are often required to work to short term budgetary constraints. This has led to an increase in competitive tendering for commercial *in-situ* testing work in recent years. Considerable reliance is thus placed upon the integrity of the company undertaking the testing, and it is unlikely that this approach will yield either optimum value for money or the greatest benefits that are potentially available from a well organised programme of *in-situ* testing.

REFERENCES

1. Concrete Society, *Concrete Core Testing for Strength*, Tech. Rep. 11, London (1976).
2. Bungey, J. H., *The Testing of Concrete in Structures*, Surrey University Press, Glasgow (1982).
3. Dept of Transport, *Inspections and Records of Highway Structures*, Departmental Standard BD 22/84, London (1984).
4. Dept of Transport, *The Assessment of Highway Bridges and Structures*, Departmental Standard BS 21/84, London (1984).
5. Institution of Structural Engineers, *Appraisal of Existing Structures*, London (1980).
6. George, A. B., Equipment and technologies in bridge inspection, *Proc. PTRC, Summer Annual Meeting*, R3, PTRC, London (1980).
7. Pullen, D. A. W. and Clayton, R. F., The radiography of Swaythling Bridge, *Atom*, (301), UKAEA, London, 3–8 (1981).
8. Danish Ministry of Transport, Road Directorate Bridge Division, *Concrete Durability*, Copenhagen (1980).

9. British Standards Institution, BS 1881 Part 201, *Testing concrete—Guide to the use of non-destructive methods of test for hardened concrete*, London (1986).
10. American Concrete Institute, *Recommended practice for evaluation of strength test results of concrete*, Standard 214–77, Detroit (1977).
11. Hindo, K. R. and Bergstrom, W. R., Statistical evaluation of the in-place compressive strength of concrete, *Concrete International*, **7**(2), 44–8 (1985).
12. Roper, H., Baweja, D. and Kirkby, G., *Towards a quantitative measure of durability of structural concrete members*, SP 82-32, American Concrete Institute, Detroit (1984).
13. Samarin, A. and Dhir, R. K., *Determination of in-situ concrete strength: rapidly and confidently by non-destructive testing*, SP 82, American Concrete Institute, Detroit, 77–94 (1984).
14. Di Leo, A., Pascale, G. and Viola, E., Assessment of concrete strength by means of non-destructive methods, *IABSE Venice Symposium*, 101–8 (1983).
15. British Standards Institution, BS 6089, *Guide to assessment of concrete strength in existing structures*, London (1981).
16. Cope, R. J., Bungey, J. H. and Rao, P. V., Assessment of reinforced concrete bridge slabs, *Proc. Int. Conf. on Rehabilitation of Structures*, IIT, Bombay, 11/103–11/108 (1981).
17. Fédération Internationale de la Précontrainte, *Inspection and maintenance of reinforced and prestressed concrete structures*. Thomas Telford Ltd, London (1986).
18. Andrews, P., *BRAINS—Bridge Record and Maintenance*, Palladian Pub. Ltd, London, **2**(6), 17–21 (1986).

2

Assessment of Concrete in Bridge Structures

J. H. BUNGEY

Department of Civil Engineering,
University of Liverpool, UK

2.1 INTRODUCTION

It is estimated[1] that there are some 50 000 concrete bridges in the United Kingdom alone, of which most have been constructed in the last 60 years. During this period there have been substantial changes in traffic flow and loading for highways so that many bridges are taking loads for which they were not designed. Characteristics of both concrete and steels used for reinforcement or prestressing have also changed, as well as design methods and site construction procedures. A further significant feature is the introduction in the last 20 years of the use of de-icing salts during cold weather. Consequently, there is a growing need for some form of assessment to be applied to concrete bridges. Most commonly, assessment is associated with identifying the nature and extent of observed or suspected deterioration. This may possibly be as a result of corrosion of embedded steel reinforcement and prestressing wire, or perhaps due to inadequacies of the concrete itself.

Assessment of the potential durability, and hence the need for future maintenance and repair, is also another aim of testing which is increasing in importance. Other circumstances include assessment of structural capacity and integrity following possible overstressing due to excessive service loads, impact damage, substandard construction or defective bearings or expansion joints. Investigations of displaced or omitted reinforcing bars, or grouting efficiency, prior to demolition of post-tensioned construction may also be important. Testing of the *in-situ* concrete may sometimes be necessary or advantageous during construction for a variety of reasons which include checking of specification compliance and timing of operations such as post-tensioning, formwork or prop removal, and curing termination.

The range of available test methods is large and includes *in-situ* non-destructive tests upon the actual structure as well as physical, chemical and petrographic tests upon samples removed from the structure, and load testing. Most techniques involve the use of modern technology but are based upon concepts developed many years ago.

Table 2.1 summarises basic characteristics of the most widely established

TABLE 2.1
Basic Characteristics of Principal Test Methods

Property under investigation	Test	Equipment type
Corrosion of embedded steel	Half-cell potential	Electrical
	Resistivity	Electrical
	Cover depth	Electromagnetic
	Carbonation depth	Chemical and microscopic
	Chloride penetration	Chemical and microscopic
Concrete quality, durability and deterioration	Rebound hammer	Mechanical
	Ultrasonic pulse velocity	Electronic
	Radiography	Radioactive
	Radiometry	Radioactive
	Permeability	Hydraulic
	Absorption	Hydraulic
	Petrographic	Microscopic
	Sulphate content	Chemical
	Expansion	Mechanical
	Air content	Microscopic
	Cement type and content	Chemical and microscopic
Concrete strength	Cores	Mechanical
	Pull-out	Mechanical
	Pull-off	Mechanical
	Break-off	Mechanical
	Internal fracture	Mechanical
	Penetration resistance	Mechanical
Integrity and structural performance	Tapping	Mechanical
	Pulse-echo	Mechanical/electronic
	Dynamic response	Mechanical/electronic
	Thermography	Infra-red
	Strain or crack measurement	Optical/mechanical/electrical

test methods classified according to the features which may be assessed most reliably in each case.

A considerable number of commercial testing organisations now offer facilities to undertake concrete inspection and testing work, and it is hoped that this chapter may provide bridge engineers who are not testing specialists with useful background information about the capabilities and limitations of the many options currently available and their relevance to particular situations requiring investigation. It must be recognised, however, that test developments and improvements of application are continuing constantly.

General guidance concerning the planning of assessment programmes, factors influencing the selection of test methods and the interpretation of results of tests on concrete structures is given in Chapter 1. This should be considered in conjunction with the more detailed information about particular test methods and their applications contained in this chapter. Space does not permit more than a brief discussion of each method, but references to books and research papers have been included to permit a more thorough study of individual techniques if this should prove necessary.

2.2 RANGE OF TECHNIQUES AVAILABLE

Available approaches may be divided into the following categories.

2.2.1 Visual Inspection

The importance of proper visual inspection has been emphasised in Chapter 1 in relation to the choice of test methods and planning of investigations. Examination of crack patterns may yield much valuable information, and reference should be made to Concrete Society Report No. 22[2] for a thorough treatment of this aspect of investigations and guidance upon the identification of crack types. A well-trained eye which knows where to look should be able to differentiate between cracking or spalling due to plastic settlement, shrinkage, structural action, reinforcement corrosion and deterioration of the concrete. Differentiation between causes of deterioration of concrete cannot normally be achieved with certainty by visual inspection alone, but the most appropriate identification tests may be selected on this basis. Pollock, Kay and Fookes[3] have suggested that systematic 'crack mapping' over a period of time is a valuable diagnostic exercise when determining the causes and progression of deterioration, and they also give further detailed guidance concerning recognition of crack types. The general nature of symptoms relating to the most common sources of deterioration have also been summarised by Higgins,[4] and Table 2.2 is based upon his suggestions. Surface crack widths may be measured with the aid of simple scales or crack microscopes, whilst a variety of calibrated 'tell-tale' devices as well as strain gauges are available to monitor movement at cracks.

Visual inspection is not confined to the concrete surface, but may include examination of bearings, expansion joints, drainage channels and similar

TABLE 2.2
Diagnosis of Defects and Deterioration

Cause of defect or deterioration	Symptoms				Age of first appearance	
	Cracking	Spalling	Erosion	Deflection	Early	Long-term
Structural deficiency	×	×		×	×	×
Reinforcement corrosion	×	×				×
Chemical attack	×	×	×			×
Alkali–silica reactions	×	×				×
Frost attack	×	×	×		×	×
Fire damage	×	×		×	×	
Thermal effects	×	×		×	×	×
Shrinkage	×			×	×	×
Creep	×	×		×		×
Rapid drying	×				×	
Plastic settlement	×				×	
Physical damage	×	×	×	×	×	×

features of a structure. Binoculars or telescopes may be useful where access is difficult, whilst borescopes are growing in usage. These small diameter optical fibre devices have their own light source and are particularly suitable for insertion into small gaps or drilled holes to permit inspection of inaccessible areas. Such locations include the interior of box sections and ducts of post-tensioned prestressed concrete, as well as bearings and expansion joints.

Visual inspection of bridges, on a regular basis, generally provides the initial indicator of the existence of some feature requiring further investigation, and when its role in planning is also taken into account this may be considered the single most important component of bridge inspection and testing.

2.2.2 Non-destructive Methods

'Non-destructive testing' is generally defined as that which does not impair the intended performance of the element or member under investigation. This, however, is usually taken in its broadest sense when applied to concrete to include methods which cause localised surface damage whilst being non-destructive in relation to the body of concrete under examination. Several of the techniques commonly used for testing concrete are of this type, and are sometimes classified as 'partially-destructive' tests, whilst other methods may leave surface marking and staining. Although the definition encompasses partially-destructive methods, techniques which require removal of a sample for subsequent testing or analysis are not commonly classified as non-destructive.

British Standard BS 1881 Part 201, 'Guide to the use of non-destructive methods of test for hardened concrete',[5] which was published early in 1986, contains outline descriptions of more than twenty methods with an indication of advantages, principal limitations and applications. There are also several other methods which are at early stages of development and have not been included in that document. Methods which have been established for some time include the use of electromagnetic cover measurement, strain gauges and radiography. These were covered by BS 4408[6] Parts 1, 2 and 3, respectively, first published in 1969 and 1970. Surface hardness and ultrasonic pulse velocity methods followed in 1971 and 1974 as Parts 4 and 5 of BS 4408, whilst the initial surface absorption test was covered by BS 1881 Part 5[7] in 1970. Revisions of all of these have recently been undertaken to take account of developments in equipment and experience, and are to be published in 1986 and 1987 as separate parts of BS 1881, 'Testing Concrete', to form a '200' series:

Part 202 Surface hardness[8]
Part 203 Ultrasonic pulse velocity[9]
Part 204 Electromagnetic cover measurement[10]
Part 205 Radiography[11]
Part 206 Strain measurements[12]
Part 207 will deal with near-to-surface partially-destructive methods and is currently under preparation
Part 208 Initial surface absorption[13]

The most established and widely used 'non-destructive' tests are described in detail by the author elsewhere[14] and in a large number of specialist research publications.

The range of properties and features that can now be assessed non-destructively is large. Tests within this category may be of value in a great many situations where investigations of concrete bridges are required relating to either material properties or local integrity. It must be remembered, however, that the results of any tests which are performed on concrete forming part of an existing structure generally relate only to the locations actually tested, and extrapolation is a matter of engineering judgement. Frequently, the accuracy with which the desired property can be established is not precise, and in some cases specific calibrations are required to relate the measured value to the required parameter, thereby introducing further uncertainties. Situations may often arise where the speed and relatively low cost of non-destructive methods may best be utilised by adopting test combinations to increase the confidence in results, as indicated in Chapter 1. In other circumstances, non-destructive tests are a useful preliminary to more expensive or disruptive methods which give a more accurate indication of the required property.

2.2.3 Methods Requiring Sample Extraction

Cores drilled from the *in-situ* concrete using a rotary diamond-headed cutting tool form the principal sample type. These may be of various sizes and used for compressive strength testing in accordance with BS 1881 Part 120,[15] or for tensile testing, determination of static or dynamic modulus of elasticity, density, water absorption, drying shrinkage and wetting expansion tests.

Apart from these physical tests, one of the most valuable uses of cores is to provide a visual indication of the internal characteristics of concrete, including aggregate distribution, cracking or honeycombing. Cores also provide convenient samples for petrographic analysis to yield information about aggregate type and characteristics, entrained air content and the presence of deleterious materials, and internal deterioration.

In some cases cores may be sprayed with an indicator immediately upon removal from the structure to indicate the depth of carbonation, which will be significant with respect to possible corrosion of reinforcement. A wide range of other chemical tests may also be performed on portions of cores to determine, for example, mix proportions, cement type, chloride content, sulphate content and the presence of admixtures. Some chemical tests may be performed on smaller drilled powdered samples taken directly from the structure, thus causing substantially less damage than that produced by coring, but the likelihood of sample contamination is increased and precision may be reduced.

As with non-destructive methods, tests on samples extracted from the structure can only reflect the conditions at the test position.

2.2.4 Tests of Structural Performance

Occasionally it may be necessary to examine the overall behaviour of an entire bridge structure or section of a bridge. This may be achieved electronically by measuring the response to dynamic loading with the aid of appropriately positioned accelerometers, or alternatively monitoring the performance under static test loads. A number of commercially available tests of dynamic performance are available, although published accounts of these are few and techniques are still under development. Static full-scale load tests, with performance based on measurement of strains and deflections together with their recovery, have, however, been in use for many years and may make use of measurement techniques described in BS 1881 Part 206[12] or precise levelling and laser methods. Large-scale tests of either of these types are likely to prove expensive but may yield valuable information concerning the 'general health' of a structure. In Switzerland, static load tests are an established component of acceptance criteria, and useful information concerning testing and deflection measurement procedures has been given by Ladner.[16,17] Such tests may also be useful in verifying analytical models, monitoring the progression of deterioration, or for assessing the effect of repairs upon overall stiffness. Detailed discussion of such testing is outside the scope of this chapter, but further information is provided in Chapter 9 relating to prestressed concrete bridges.

2.3   ASSESSMENT DURING CONSTRUCTION

There is an increasing awareness of the possible benefits of *in-situ* testing of concrete during construction, which are principally related to long-term durability. Physical checking of the thickness of concrete cover above reinforcing bars to ensure that they have not been moved during concreting, for example, is achieved simply with the aid of electromagnetic devices.[10] Since this parameter is recognised as being a major factor in preventing future corrosion of reinforcement, initial 'acceptance' checking would clearly provide long-term benefits but is often not adopted because of the extra initial cost, although this would usually be minimal in relation to the overall costs of a new project.

A similar argument based on cost is often used against routine *in-situ* strength testing, and reliance is still placed upon standard 'control specimens' of the concrete tested in a laboratory. Even where a concrete strength value is required before formwork stripping, prop removal or post-tensioning, cubes or cylinders which have been stored alongside the structure are still commonly used despite the wide discrepancies which are known to be likely to exist between these and the true strength of the concrete in a structural element.

Pull-out tests, in which the force required to pull out an insert cast into the concrete surface is measured, and penetration resistance tests, in which the depth of penetration of a probe fired into the concrete surface is measured, are both *in-situ* testing techniques which have gained acceptance in several parts of the world. Pull-out tests in particular are aimed specifically at assessing the development of *in-situ* strength during construction, and their use is slowly increasing.

It should also be noted that in Scandinavia there is an increasing usage of thin-section microscopy of samples obtained from cores extracted from newly constructed concrete to ensure that cement type and content are as specified. Assessment of surface absorption characteristics is a further approach which is sometimes employed as a measure of the surface quality, finish and potential durability, but usage in bridge construction has again been very limited. These approaches are all discussed in greater detail in subsequent sections of this chapter.

2.4   ASSESSMENT OF DETERIORATION

Deterioration of reinforced and prestressed concrete bridges may be divided into that caused by chemical and physical environmental effects upon the concrete itself, and damage resulting from corrosion of embedded steel.

2.4.1 Sulphate Attack

One of the principal chemical actions encountered with Portland cement concretes is sulphate attack, which causes an expansive reaction and disruption of the concrete. This is particularly associated with ground waters containing dissolved aggressive salts and may present a major problem in areas such as the Middle East. Soft moorland water presents the principal such source within the United Kingdom, and if the water is acidic with a pH value below 6·0 the situation is likely to be particularly severe. Where external attack is caused by ground water, problems are most likely to occur in buried concrete and are thus not readily detectable, although deterioration of wing walls may provide a useful pointer. Other sources of sulphate attack include the use of contaminated aggregates leading to internal loss of strength and disintegration, and attack of exposed surfaces resulting from atmospheric pollution. External attack from wind-blown dust containing high concentrations of sulphates, in conjunction with night-time condensation, is a further problem in some areas of the world. External attack will usually be characterised by irregular surface cracking within a few years, followed by disintegration and erosion.

The principal method of identifying sulphate attack is to measure the sulphate content of a sample of concrete by chemical analysis. If the sulphate content exceeds about 3% of the cement content for concretes made with Portland cements, chemical attack may be indicated. It is thus recommended that cement content determination is performed in conjunction with sulphate tests, and a sample in the form of an intact piece of concrete is preferable. Detailed procedures are given in BS 1881 Part 6,[18] together with guidance on sampling procedures and precision. It is particularly important to recognise that, as with any chemical analysis, results will relate only to the particular location from which the sample was obtained and it is important that sufficient samples are taken to be representative.

Sulphate attack may also be identified by petrographic methods, as described by Power and Hammersley,[19] which involve the microscopic study of thin sections to reveal the presence of crystalline calcium sulphoaluminate. This is commonly known as ettringite, and may be found along cracks within the concrete when chemical reactions between sulphate bearing solutions and the cement paste have occurred. Whilst this is a powerful technique, unfortunately the cost of sample preparation is high and the use of this approach is likely to be inhibited for this reason. Other specialised chemical techniques, including differential thermal analysis, may also be applied to sulphate identification, but basic chemical analysis will be preferred for most practical situations where an appropriate sample can be obtained.

2.4.2 Alkali–Aggregate Reactions

Although common in some parts of the world, including Canada and Scandinavia, this problem was thought initially to be localised within the United Kingdom, but events have subsequently shown that the problem may potentially be widespread across the country. Although concrete in the Middle East may be subject to deterioration from a variety of causes, including materials deficiencies, alkali–aggregate reactions are not generally considered to be a major problem in that region of the world. This topic has been the subject of considerable publicity in the UK in recent years, following initial reports of deterioration of a viaduct near Plymouth in 1980.[20]

The problem may be encountered when cements with a particularly high alkalinity are used with certain reactive aggregates. In the presence of moisture, an expansive internal reaction may take place, causing disintegration of the concrete matrix. Reactions may be of a variety of forms, of which the most common are alkali–silica, alkali–silicate and alkali–carbonate. In the United Kingdom, the problem appears to be confined to reactions between alkalis and silica in microcrystalline and cryptocrystalline form and chalcedony as found in flints and cherts. The latter may sometimes be present in limestones, whilst strained quartz in some quartzites may also give trouble. Such reactions are commonly referred to by the term alkali–silica reaction (ASR) in the UK, where in most reported cases the reactive constituents have been found to be present in the fine aggregates, although there have been some cases involving the coarse aggregates.

A number of reports have recently been published with a view to minimising the risk of such reaction in new structures, but very little published guidance is at present available concerning testing and diagnosis, although the Cement and Concrete Association is currently preparing a report on this topic. The most comprehensive guidance presently available is given by Wood,[21] who emphasises the importance of not overlooking other sources of cracking[2] before embarking upon detailed testing. 'Map-cracking', illustrated in Fig. 2.1 (photo by courtesy of Professor R. J. Cope), is commonly associated with ASR, although Wood states that this is only apparent in unstressed unreinforced zones or when the reaction has reached an advanced stage. At early stages ASR will tend to exaggerate and promote the growth of crack patterns associated with normal structural and material behaviour as the gel produced by the reaction expands into existing cracks. Visual observation of gel 'oozing' from surface cracks may be a useful indicator of the presence of the reactions, whilst reactive aggregate may also lead to surface 'pop-outs'. Cracking seldom appears in under five years, but in extreme cases crack growth rates may subsequently be of up to 1·0 mm/year.

The presence of moisture is essential for such reactions to occur and a critical value of 75% relative humidity has been quoted[22] below which the reaction will cease, but remain dormant to recommence if the moisture level increases above this value. Moisture variation will thus augment the

Fig. 2.1.   Photograph of typical alkali–silica reaction map-cracking.

materials variations that are always present within *in-situ* concrete to lead to very high variability of behaviour of field concrete containing potentially reactive ingredients. This will be of great significance when carrying out investigations, and in particular the likelihood of increased reaction below ground level due to greater dampness levels must be recognised. Evidence also suggests that reactions are likely to be aggravated by de-icing salts, rendering highway bridge structures particularly vulnerable, especially in zones subject to spray or leakage due to inadequate drainage or joint performance.

There are currently no standardised test procedures for identifying or classifying these reactions, and although ASTM C227-81[23] and C289-81[24] relate to potential reactivity of aggregates and aggregate cement combinations and are commonly used on a worldwide basis, it is claimed[22] that neither permits effective classification of British aggregates. Procedures currently in use in the UK to identify and classify ASR place emphasis upon expansion tests on cores extracted from the *in-situ* concrete. It is recommended[21] that a 75 mm diameter core with 9 Demec gauge spans of 50 mm is appropriate, with the reaction being accelerated for diagnosis purposes by a 38°C, 100% relative humidity environment. Expansion measurements may be coupled with weight change and ultrasonic pulse velocity measurements on the cores, and by chemical analysis of representative samples for alkali and chloride content. Petrographic methods involving thin sections will usually be necessary to identify the reactive minerals, as shown in Fig. 2.2. In this particular case an opaline chert particle in the centre of the photomicrograph shows a typical ASR crack pattern. The gel-coated air voids and cracks in the surrounding cement paste are clearly visible. Inert quartz grains are black as they are impervious to the fluorescent epoxy resin. This approach may be coupled

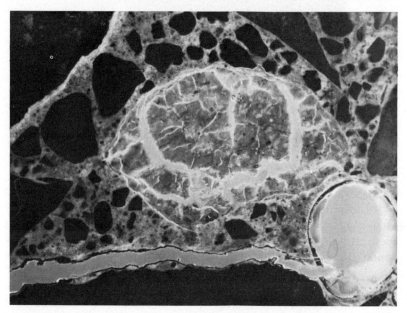

Fig. 2.2.   Photomicrograph of alkali–silica reaction (reproduced by courtesy of AEC Consulting Engineers, Holte Midtpunkt 23-3, 2840 Holte, Denmark).

with observation of the degree of cracking and pulse velocity measurements to give some indication of the extent of the reaction completed, although this cannot yet be quantified. A useful indication of the extent of internal cracking may also be obtained by observation under ultra-violet light of a cut surface impregnated with fluorescent dye.

Classification in the UK is thus currently based upon the degree of latent expansion indicated by three-month values of accelerated tests,[21] in conjunction with all the other test results, to permit assessment of likely future serviceability problems and requirements for maintenance and remedial works. Long-term expansion tests at more realistic temperatures and relative humidities may be worthwhile for structure management purposes, coupled with *in-situ* measurements of surface crack widths and their rate of development. Assessment of the future *in-situ* performance and the monitoring of possible remedial measures require the development of a reliable procedure for the measurement of *in-situ* moisture conditions. One possible approach to this is to measure the moisture content of a small timber specimen which has been sealed into a drilled hole for a sufficient period to reach moisture equilibrium with the surrounding concrete. A chemically-based humidity meter (HUM meter) which is inserted into a surface drilled hole is also commercially available from Denmark, although little published evidence is available concerning its accuracy.

Strength assessment of concrete affected by this reaction is complex, and as yet no reliable procedure is available which takes account of the confinement effect provided by surrounding concrete and reinforcement to the interior of a body of microcracked concrete. There will clearly be a loss of tensile, bond and shear strength, but it is generally considered likely that where there is a good, well-detailed three-dimensional cage of reinforcement, overall strength effects may not be serious. Where shear steel or links are not provided, however, the risk of delamination may lead to serious loss of shear and flexural capacity. Whilst pull-out and pull-off test methods may indicate surface zone strength characteristics, these are likely to be of little value unless they can be related comparatively to general behaviour characteristics, and at the present time this is not possible.

2.4.3 Freeze–Thaw Damage

This action, which is aggravated by the presence of de-icing salts, is one of the principal physical actions likely to affect the performance of highway structures and may lead to surface scabbing and long-term disintegration. Although it is normal practice in the United Kingdom to specify air-entrained concrete for structural components which are likely to be exposed to such attack, cases may still occur particularly as a result of spray. It is virtually impossible to distinguish reliably between freeze–thaw action and alkali–aggregate reactions without the use of thin-section petrographic techniques. Assessment of the entrained air content of hardened concrete may also be obtained by a similar approach, although the more usually adopted procedure is that recommended by ASTM C457-80,[25] in which air bubbles are counted by stereo microscopic examination of a polished, dyed, sawn concrete surface. This approach has been discussed by Bungey.[14] Samples for either type of examination may conveniently be obtained from the structure in the form of cores.

2.4.4 Fire Damage
Whilst this is relatively uncommon for bridge structures, damage may range from localised surface spalling to major structural deterioration if exposure has been sufficient to permit significant temperature rises in reinforcement or prestressing steel. Visual assessment of cracking, spalling and colour change is likely to provide the major source of information upon which an assessment can be based. Concrete Society Technical Report 15[26] provides useful guidance, although ultrasonic pulse velocity and thermoluminescence techniques may provide additional data in some circumstances.[14] Data relating to the performance of steel and concrete under fire conditions has also recently been included in Part 2 of BS 8110.[27]

2.5 CORROSION OF REINFORCEMENT AND PRESTRESSING STEEL

Corrosion of embedded steel is the major cause of deterioration of concrete bridges in the UK at the present time. This may lead to structural weakening due to loss of cross-section of reinforcing bars or prestressing wires, surface staining and cracking or spalling, or internal delamination of bridge decks in which a concrete fracture plane occurs at the level of the corroding reinforcing steel mesh. Testing will thus be directed towards identifying the risk of corrosion which is not apparent at the surface, the extent of corrosion which is partially observable at the surface, and the integrity of a structure which may have experienced corrosion.

2.5.1 Assessment of Risk and Extent of Corrosion
Steel corrosion is an electrochemical process involving the establishment of anodic (corroding) and cathodic (passive) sites on the metal, leading to the

development of regions of differing electrical potential, as illustrated in Fig. 2.3.

Corrosion results in a pattern of current flow within the concrete reflecting these electrochemical potentials, provided that oxygen and water are available in sufficient quantities at both sites to fuel the process. Measurement of the potentials and the magnitude of corrosion currents may therefore, *in principle*, enable an assessment of the existing and future corrosion behaviour to be made. *In practice*, concrete has cracks, flaws and aggregate particles which affect diffusion flow characteristics, and it is not in uniform contact with the reinforcement. As a result, measurements of electrical characteristics can, at present, provide only a very general guide

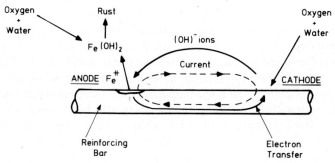

Fig. 2.3.   *Basic corrosion process of reinforcing bar in concrete.*

to corrosion conditions, although they are the only practicable methods currently available which can offer any real help in locating corrosion zones.[28]

Corrosion of carbon steel will not normally occur in a highly alkaline environment such as that provided by the pore fluid in hardened concrete. It is only when the passivation provided by the surrounding concrete is broken down—most notably by carbonation and the presence of chlorides—that corrosion may commence as either micro- or macro-corrosion cells. Macro-cells are likely to be the most disruptive, due to expansion of the rusting steel, although micro-cells (which are much more difficult to detect) may locally severely reduce the bar cross-section with little external evidence.[29]

Two principal electrical test methods are currently in use,[30] although basic research is still in progress relating to both, and further developments may be expected.

2.5.1.1 Half-cell Potential Measurements

This method has been widely used with success in recent years where reinforcement corrosion is suspected, and normally involves measuring the potential of an embedded reinforcing bar relative to a reference half-cell placed on the concrete surface, as shown in Fig. 2.4. This is usually a copper/copper sulphate or silver/silver chloride cell but other combinations have been used. The concrete functions as an electrolyte, and the risk of corrosion of the reinforcement in the immediate region of the test location may be related empirically to the measured potential difference.

The basic equipment is simple and permits a non-destructive survey to produce isopotential contour maps of the surface of a concrete member, as in Fig. 2.5. Zones of varying degrees of corrosion risk may be identified from these maps on the basis of the information in Table 2.3 coupled with a study of localised potential rates of change, although the method cannot indicate the actual corrosion rate. Localised corrosion risk is indicated by 'whirlpool' effects, generalised corrosion by more uniformly low potentials.

A small hole will probably need to be drilled to enable electrical contact to be made with the reinforcement in the member under examination (this connection is critical, and a self-tapping screw is recommended), and some surface preparation, including wetting, may also be required. Commercially available equipment includes 'multi-cell' and 'wheel' devices with automatic data logging facilities, both of which are designed to speed site measurements sufficiently to permit large areas of concrete surfaces to be tested economically.

2.5.1.2 Resistivity Measurements

The electrolytic resistivity of concrete is known to be influenced by many factors, including moisture and salt content and temperature, as well as mix

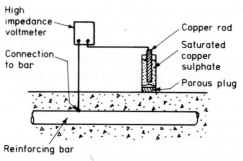

Fig. 2.4.   Half-cell potential test.

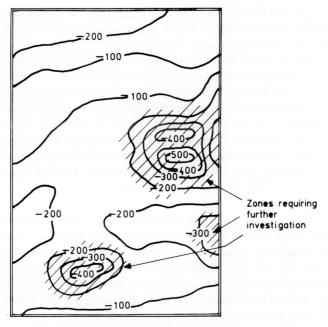

Fig. 2.5. Half-cell potential contours.

TABLE 2.3
Interpretation of Electrical Test Results

Half-cell potential (mV) relative to copper/copper sulphate reference electrode	Percentage chance of active corrosion
< -350	95
-200 to -350	50
> -200	5

Resistivity (ohm-cm)	Likelihood of significant corrosion (non-saturated concrete when steel activated)
$< 5\,000$	Almost certain
$5\,000-12\,000$	Probable
$> 12\,000$	Unlikely

proportions and water/cement ratio. The presence of reinforcement may also influence measurements. *In-situ* measurements may be made using a Wenner four-probe technique, in which four electrodes are placed in a straight line on, or just inside, the concrete surface at equal spacings. A low-frequency electrical current is passed through the outer electrodes whilst the voltage drop between the inner electrodes is measured, as shown in Fig. 2.6. The apparent resistivity of the concrete, and hence ease of current flow within the concrete, may be calculated from a knowledge of the current, voltage drop and electrode spacing. For practical purposes, the depth of the zone of concrete affecting the measurement may be taken as equal to the electrode spacing, but other practical influences, such as test member dimensions and efficiency of surface coupling, may affect the measured values.

The technique can provide a simple non-destructive indication of the electrical resistivity of the concrete at the test location. This can be related, principally by experience, to the corrosion hazard of embedded reinforcement, and the likelihood of corrosion may be predicted in situations in which half-cell potential measurements show that corrosion is possible (Table 2.3). At the present time, this method is less widely used in the UK than half-cell potential measurements.

Other tests which may also provide useful information about the risks and causes of reinforcement corrosion involve measurement of the extent of carbonation and chloride penetration below the concrete surface.

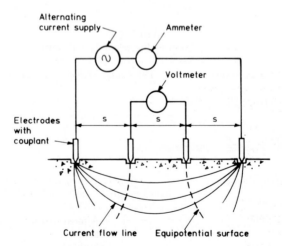

Fig. 2.6.   Apparent resistivity measurement.

2.5.2 Carbonation Depth Measurement

The loss of alkalinity associated with carbonation of surface zone concrete may be detected by the use of a suitable indicator (e.g. phenolphthalein) which can be sprayed on to a freshly exposed cut or broken surface of the concrete. If phenolphthalein is used a purple–red coloration will be obtained where the highly alkaline concrete has been unaffected by carbonation, but no coloration will appear in carbonated zones. This is a quick and simple approach which gives an immediate visual indication of the position of the depassivation front relative to the steel. Other variations of this approach include the use of a 'rainbow' indicator[31] to give a more detailed indication of alkalinity levels, whilst the localised effects of cracks can be examined by microscopic examination of thin sections.

A knowledge of the depth of carbonation combined with the age of the concrete may be useful when assessing potential durability and the likelihood of future corrosion of reinforcement. Features of carbonation have been discussed in detail by Roberts,[32] and a considerable amount has recently been written about this topic. Somerville[33] summarises aspects of the penetration of carbonation with time, and the ensuing significance in terms of long-term durability, identifying a two-phase mechanism for corrosion of reinforcement, as illustrated in Fig. 2.7.

The rate of penetration of carbonation will depend on many factors, including those associated with the mix such as water/cement ratio, those associated with construction such as curing, and environmental conditions.

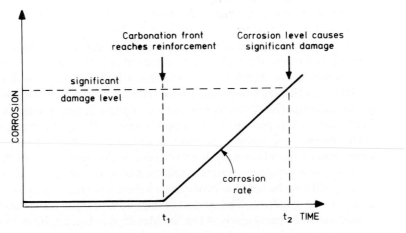

Fig. 2.7.   Influence of carbonation on reinforcement corrosion.

It is not possible at present to predict carbonation rates with any degree of confidence, although results are available for some specific combinations of these variables. It is generally accepted, however, that the progression under particular circumstances may be obtained from the expression

$$D = K\sqrt{t}$$

where D = depth in mm and t = age in years, whilst K is a constant for the actual concrete. Future predictions may thus be made on the basis of a carbonation depth measurement at known age. It must be remembered, however, that rates may vary across a member or structure according to localised variations in atmospheric exposure conditions. Somerville[33] suggests that the most severe condition is likely to be associated with a relative humidity of 60–70%, as may be expected with external concrete sheltered from the weather.

2.5.3 Chloride Ion Penetration

This may be identified by chemical analysis of powdered drillings obtained from various depths below the concrete surface or from samples taken at varying positions of a core sample, and may similarly provide useful information about the likelihood of depassivation of the reinforcing steel.

An important feature of chloride ions is their mobility within concrete in the presence of moisture. De-icing salts represent the most likely source of chlorides in bridge structures and, as with carbonation, a knowledge of the rate of penetration from the surface is valuable when predicting the likelihood of future reinforcement corrosion. A model similar to that shown in Fig. 2.7 may also be used in this respect, with a level of 0·35% chloride ion by weight of cement sometimes considered to be sufficient to promote corrosion, although it may be present at lower values.[29] Chloride measurements are therefore not suitable for locating corrosion but are useful in establishing causes and appropriate remedial action. This may be particularly significant in situations where sulphate-resisting cements have been used, as may perhaps be the case for wing walls or abutments, since such cements are known to have a reduced resistance to chloride penetration. Chloride ions are, however, very small in relation to the typical pore sizes of good quality concrete, and most concretes are thus vulnerable. Testing will thus be aimed at providing a 'chloride profile' within the cover zone of the concrete. The zones of highway bridges which may be considered to be particularly at risk are identified in Fig. 2.8, from which it is apparent that few parts are likely to be free from exposure to chlorides.

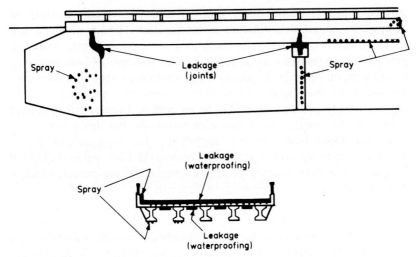

Fig. 2.8.   Vulnerability of highway bridges to chlorides.

Simple site tests[14] such as the 'Hach' and 'Quantab' methods may quickly provide useful preliminary information in most cases prior to more thorough laboratory tests, if these are needed. The recommended laboratory method is the 'Volhard' approach which is relatively straight-forward and reliable, whilst x-ray fluorescence spectrometry will yield both chloride and cement content simply and quickly, although requiring specialised sample preparation and test equipment. Another newly developed simple approach, involving observations of colour differentials on the surface of a core sprayed with a chemical indicator, is also available from Denmark.[31]

Results will frequently be expressed as per cent chloride ion by weight of concrete which, assuming a cement content within the range 11–20%, will often be of sufficient accuracy for general surveys, but in cases of dispute a precise cement content determination will normally also be required.

2.6   ASSESSMENT OF STRUCTURAL INTEGRITY

Localised loss of structural integrity in bridges is most likely to result from delamination caused by corrosion expansion of the outer layer of reinforcement, particularly where the cover is low.

Experience has shown that the human ear, used in conjunction with surface tapping, is the most efficient and economical method of determining major delamination in bridge decks. Chain dragging is a variation on this which is commonly used in North America, where the problem is widespread as a result of the absence of waterproofing membranes on older bridge decks. Such approaches rely on a subjective assessment by the operator to differentiate between sound and unsound regions, and results cannot readily be quantified. Various attempts have been made to monitor the response electronically, but this is not simple since factors such as uniformity of applied blow become significant. This is an area in which research is proceeding both in the USA and in the United Kingdom. The most recently produced commercial equipment, which is called an instrumented delamination device, is hand-held and provides a digital measure of the amplitude of the reflected wave, using the pulse-echo technique as illustrated in Fig. 2.9. Whilst this equipment is simple to use, the amount of information that can be obtained is limited, and application is restricted to comparative surveys to determine major differences in the thickness of sound concrete. A more sophisticated approach involves the use of an accelerometer coupled to an oscilloscope, which permits a detailed study of the characteristics of the reflected sound waves. Such equipment would probably require to be vehicle-mounted for use on site.

A plot of surface temperature contours obtained by an infra-red thermographic survey may also be used to identify sound and delaminated areas of bridge decks[34] as well as hidden voids and ducts. Surface temperature differentials will exist during heating and cooling as a result of variations in internal heat transfer characteristics. Unfortunately these differences are very small and, although surveys may be conducted from

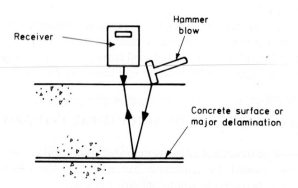

Fig. 2.9.   Pulse-echo method.

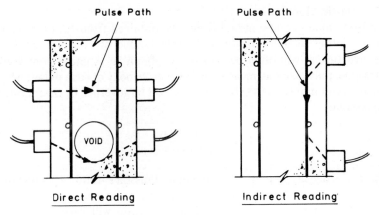

Pulse Path   Pulse Path

Direct Reading   Indirect Reading

Fig. 2.10.   Ultrasonic pulse velocity measurements.

helicopters or similar vantage points while the deck is still in use and may penetrate through overlays, great care is needed to avoid extraneous effects, and the equipment is very expensive.

Techniques are also available to locate reinforcing bars using this approach, with the assistance of electromagnetic induction heating, but applications of this to bridges are likely to be limited.

Through-transmission ultrasonic methods can be used to locate substantial air-filled cracks or voids, as illustrated in Fig. 2.10. Areas of poorly compacted concrete are likely to cause attenuation of the pulses and may similarly be detected by means of a comparative survey.

Unfortunately, testing access is often restricted to one surface only, rendering this approach of little practical value, although it may be possible to estimate the depth of penetration of substantial surface cracks. Short-wave subsurface radar scanning[34] is under development in the USA for the location of voids beneath road pavements, and may have future applications to some aspects of concrete bridge construction.

2.7   ASSESSMENT OF PHYSICAL PROPERTIES OF CONCRETE

Testing of concrete properties is most likely to be required to provide data for comparison with specifications, incorporation into calculations of serviceability and loading capacity, and for the assessment of potential durability and future repair and maintenance requirements.

2.7.1 Elastic Modulus

Whilst a value of static modulus may be obtained from strain measurements during load testing of a suitable sample of concrete cut from
the structure, it will often be possible to obtain an estimate from *in-situ*
ultrasonic pulse velocity measurements. The velocity of such pulses, which
are most commonly of 54 kHz frequency, will be governed by the following
expression:

$$V = \sqrt{\frac{(1-v)}{(1+v)(1-2v)}\frac{E_d}{\rho}}$$

where $\rho =$ density (kg/m^3), $v =$ dynamic Poisson's ratio, $E_d =$ dynamic
elastic modulus (kN/mm^2) and $V =$ ultrasonic pulse velocity (km/s).

Since the density and dynamic Poisson's ratio will vary over only a
relatively small range for most concretes made with naturally occurring
normal-weight aggregates, a relatively well-defined relationship exists
between pulse velocity and dynamic elastic modulus. This may be related to
the static modulus by the use of the relationship given in BS 5400.[35] The
only major limitation to this approach is the need to establish a well-defined
pulse path free of cracks, voids and substantial reinforcement. A 'direct'
measurement between two opposite faces of concrete is preferred, since the
path length is more clearly defined and transverse reinforcement effects are
likely to be less than with 'indirect' readings, as illustrated in Fig. 2.10.[36]

2.7.2 Density

Absolute values of density may be obtained from physical testing of cores
or other suitable samples, whilst relative *in-situ* densities can be obtained by
radiography or radiometry.

2.8   ASSESSMENT OF CONCRETE STRENGTH

Whilst strength is seldom the most critical property of concrete in relation
to its long-term structural behaviour, it forms the basis of compliance with
most specifications. The strength of 'standard' cube or cylinder specimens
taken from the fresh mix at the point of placing is usually used for this
purpose and also as the basis of calculations, although it is recognised that
these specimens cannot reflect the actual compaction and curing applied to
the *in-situ* concrete. In the event that these 'standard' specimens fail to meet
the specified strength, it will usually be necessary to embark upon a

programme of non-destructive testing to permit an assessment of structural adequacy of the suspect concrete in its particular location within the structure. Such testing will not be planned in advance and is not regarded as part of routine quality control procedures. If the position and extent of the suspect concrete is unknown this may be identified by comparative surface hardness or ultrasonic measurements, but unless specifically developed calibrations are available these methods should not be relied upon for absolute strength estimates. Particular care is required with the location and interpretation of all *in-situ* strength tests, including their relevance to 'standard specimen' results, as discussed fully in Chapter 1.

An indication of the strength of the *in-situ* concrete may also be required at later stages in the life of a bridge. For example, when a strength reduction is suspected or, more commonly, when an older structure for which few records are available is being assessed for upgrading, reinstatement or repair. In these cases, tests will normally be located at positions which can be identified as critical.

The most reliable method of obtaining an estimate of the strength of concrete in a structural element is the cutting of cores for subsequent preparation and crushing in a laboratory. However, to achieve a worthwhile accuracy, these cores must be at least 100 mm in both diameter and length, and at least three are required from a given location. It is generally accepted that an estimate of actual *in-situ* cube strength is unlikely to have an accuracy better than $\pm 12/\sqrt{n}\%$, where n is the number of samples tested. If smaller diameters are used, then a much greater number of samples is required[37] to achieve comparable accuracy. BS 6089[38] provides basic guidance on interpretation of core results, including allowances for specimen shape, proportions and orientation, whilst more extensive information on this subject is provided by Concrete Society Technical Report No. 11.[39] Particular care is needed to locate cores so as to provide representative evidence and to avoid reinforcement wherever possible. It is also essential that the orientation and location of individual specimens are carefully identified.

Frequently, the damage caused and the delays while the cores are cut, prepared and tested are unacceptable. The 'partially-destructive' techniques that are available for assessing strength of surface zone concrete are generally less reliable than cores, but cause substantially less damage and give instant results. These include established methods such as penetration resistance (Windsor probe), pull-out (Lok & Capo), internal fracture, break-off and pull-off methods. A further approach using an expanding sleeve has also recently been proposed and called the ESCOT method.[40] All have the important characteristic that they directly measure a strength-related property and strength calibrations are therefore not as sensitive to such a wide range of variables as the truly non-destructive methods.

2.8.1 Penetration Resistance Test

The most commonly known form of this approach is the Windsor probe test in which a hardened steel alloy probe is fired into the concrete surface by a driver using a standardised powder cartridge. This equipment is illustrated in Fig. 2.11. The depth of penetration will usually lie between 20 and 40 mm, and is believed to be influenced by the concrete in the outer 75 mm zone. The penetration depth is measured, and the mean of three readings, related empirically to compressive strength by calibration charts, may yield an *in-situ* strength estimate with 95% confidence limits of about ±20%. The principal factors influencing this relationship are aggregate hardness and type as well as power level, and specific calibration is essential.[14] The method has been available in the USA for many years, and its use in the UK has increased slowly. Although unsuitable for slender members, the test is quick and useful where access may be difficult, especially if comparative strength values only are required.

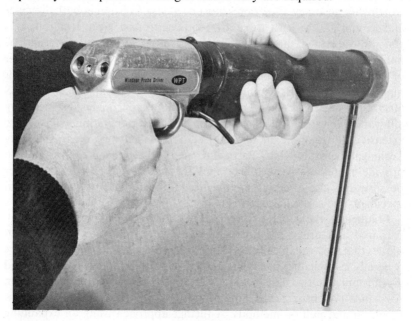

Fig. 2.11.   Windsor probe test.

2.8.2 Pull-out Tests

Developed principally in Denmark during the past 20 years, these methods are gaining acceptance in Scandinavia and the USA. A circular steel insert of 25 mm diameter is located at a depth of 25 mm below the concrete surface and pulled by means of a calibrated hand-operated hydraulic jack against a 55 mm diameter reaction ring, as shown in Figs 2.12 and 2.13. The configuration is such that failure is believed to be dominated by compression of the concrete between the insert and the reaction ring and thus relatively independent of other properties.

The pull-out test is commercially available in two forms: the Lok-test, which uses an insert cast into the concrete; and the Capo-test, in which a compressed steel ring is expanded into a groove undercut from an 18 mm diameter drilled hole. The Lok-test is intended for quality control and strength development monitoring purposes, whilst the Capo-test is intended for strength assessment of existing concrete. In both cases, the average of six readings would normally be used for correlation against compressive strength.[14] For practical purposes this relationship may be regarded as 'general' in nature for concretes made with natural aggregates, and an accuracy of *in-situ* strength prediction of the order of $\pm 20\%$ should be possible on this basis. This accuracy may be improved to about $\pm 10\%$ if a specific calibration is obtained for the mix in use.

2.8.3 Internal Fracture Test

Developed at the Building Research Establishment[41] in response to the need for a simple site test for strength measurement of high alumina cement concrete, application of the method has since been extended to concrete made with Portland cement. A 6 mm diameter hole is drilled approximately

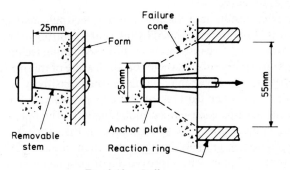

Fig. 2.12.   Pull-out test.

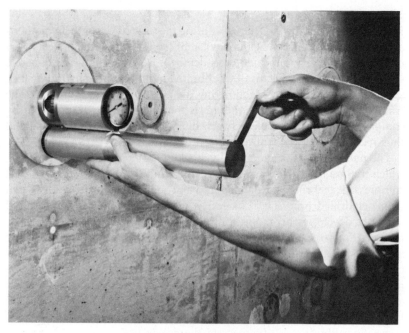

Fig. 2.13.   Lok-test equipment.

35 mm deep into the concrete surface using a masonry drill. An expanding wedge anchor bolt is fixed into this hole to a depth of 20 mm and pulled against an 80 mm diameter reaction tripod by a torque-meter acting on a greased nut, as shown in Fig. 2.14. The peak torque is observed, and the average of six such values may be related to compressive strength with the aid of calibration curve.

   The scatter of results may be lower if a direct-pull load application method is used.[14] Failure is initiated by internal fracturing of the concrete, but the load necessary to cause this is sensitive to loading technique, which must be carefully standardised. An important feature of this method is that a 'general' calibration curve may be used, since sensitivity to aggregate characteristics and other features of the concrete mix is low, but it is recommended that this is developed individually by the operator for the particular equipment and loading technique in use. Accuracy of *in-situ* strength prediction is unlikely to be better than about ±28% using a general calibration, but may be improved if a direct-pull loading technique is used with its own relevant calibration.

Fig. 2.14. Internal fracture test.

2.8.4 Pull-off Test

A circular steel probe is bonded to the surface of the concrete by an epoxy resin adhesive. A specially designed portable apparatus is then used to pull off the probe, along with a bonded mass of concrete, by applying a direct tensile force. The peak load is measured, and as the failure area is approximately equal to that of the probe, a tensile strength can be calculated, and an equivalent cube strength estimated with the aid of an appropriate calibration graph. A modified version using partial coring can be used to overcome hard surface shell effects. Careful surface preparation is necessary but the results have been found to be consistent and reliable, and a large-scale *in-situ* test programme has been undertaken with success.[42] Accuracies of strength estimation of ±15% are claimed.

2.8.5 Break-off Test

This method, which has been developed in Scandinavia, measures the force required to break off a 55 mm diameter core which has been effectively formed within the concrete to a depth of 70 mm. This may be achieved by a disposable tubular plastic sleeve inserted into freshly placed concrete, or by drilling a circular slot if existing concrete is to be tested. An enlarged slot is formed near to the surface into which a load cell coupled to a hydraulically-operated jack is inserted to provide a transverse force to the top of the core. This will cause a flexural failure at the base of the core and the break-off force may be related to compressive strength by means of appropriate calibration charts. The mean of five tests would normally be required for this purpose, although the tests are quick to perform.

2.8.6 ESCOT Test

This newly proposed technique[40] involves the use of an internally tapered sleeve which is longitudinally slotted at its tapered end and is placed over a bolt with a flared end matching the sleeve taper. The combined bolt and sleeve assembly, which is 20 mm in length and 11 mm in outside diameter, is inserted into a hole drilled into the concrete surface and fitted to a loading assembly consisting of a bearing cylinder, ball-race housing and nut. Torque is applied to the nut, causing the sleeve to expand until failure is recorded by the peak torque value. Whilst data and experience are still very limited, it is claimed that results are promising and compare favourably with internal fracture and pull-out results with regard to accuracy of strength prediction based on appropriate calibration curves.

2.9   MIX CONSTITUENTS AND PROPORTIONS

2.9.1 Cement Type

The basic type of cement used in a concrete may be established by separation and chemical analysis of the matrix, for comparison with analyses of particular cement types. Alternatively, microscopic examination can be used to detect unhydrated cement particles which can be compared with known specimens. The complexity of the analysis varies according to cement type, admixture, replacements and aggregates, and BS 1881 Part 6[18] strongly recommends that chemical analysis should be supplemented by microscopic examination. It is not possible to distinguish between ordinary and rapid-hardening Portland cements, although cement replacements can usually be detected. Complex techniques such as differential thermal analysis may also be used to establish cement type on a comparative basis.

2.9.2 Cement Replacements

The most commonly used cement replacements are granulated ground blast furnace slag and pulverised fuel ash (PFA). Blast furnace slags contain considerably higher levels of manganese and sulphides than are to be found in normal cements, and this can be used for their identification by chemical analysis. If the slag is from a single source of known composition it may be possible to obtain quantitative data, provided there is no slag in the aggregate. A characteristic green or greenish-black coloration of the interior of the concrete may aid identification, and microscopic methods may also be used. PFA is most easily detected by microscopic examination of the acid-insoluble residue of a sample of separated matrix. Particular, characteristic spherical particle shapes may be recognised, although quantitative assessment of the PFA content is not possible. Microscopic examination of thin sections, as described by Power and Hammersley,[19] may also be used.

2.9.3 Cement Content, Aggregate/Cement Ratio and Grading

The most common methods for determining the cement content of a hardened mortar or concrete are based on the fact that lime compounds and silicates in Portland cement are generally more readily decomposed by, and soluble in, dilute hydrochloric acid than the corresponding compounds in the aggregate. The quantity of soluble silica or calcium oxide is determined by simple analytical procedures and, if the composition of the cement is known, the cement content of the original volume of the sample can be calculated. Allowance must be made for any material which may be dissolved from the aggregate, and representative samples of aggregate should be analysed by identical procedures to permit corrections to be made. It will not be possible to treat all concretes in an identical manner because of differences in aggregate properties, but the aggregate/cement ratio will be found as well as cement content, whichever method is used. If gradings are required, methods involving crushing of the concrete cannot be used.

The importance of careful sampling and sample preparation cannot be overemphasised. A truly representative test specimen is more likely to be achieved by grinding than by physical separation. If the aggregates prevent analysis of cement content and aggregate/cement ratio by chemical means, it may be possible to obtain estimates based on micrometric methods or by techniques such as x-ray fluorescence spectrometry.

Whatever method is used to determine cement content, it is essential to recognise that no statistical statement about a particular batch of concrete can be made from a single sample whose test accuracy is unlikely to be better than $\pm 25 \, \text{kg/m}^3$. Where several separate samples are used, an estimate of an average value may be possible, and a minimum of four individual samples, or preferably six, should be used to obtain a reliable estimate relating to a limited volume of concrete (say $10 \, \text{m}^3$). For large volumes, or a large number of units, 10 to 20 independent random samples, each analysed separately, are recommended by Lees.[43]

2.9.4 Original Water Content

The quantity of water present in the original concrete mix can be assessed by determining the volume of the capillary pores which would be filled with water at the time of setting, and measuring the combined water present as cement hydrates. The total original water content will be given by the sum of pore water and combined water; the method is described in BS 1881 Part 6.[18] It requires a single sample of concrete which has not been damaged either physically or chemically. Usually the water/cement ratio is of greatest interest, so that a cement content determination will also be necessary.

Since this determination requires a sound specimen of concrete, strength-tested cores cannot be used. Lees[43] has suggested that a precision better than ±0·05/1 for the original water/cement ratio cannot be achieved even under ideal conditions. A major source of difficulty lies in the corrections for aggregate porosity and combined water which may be overestimated by the procedures used. This will lead to an underestimate of the true original water content, whilst the original cement content must also be measured with reasonable accuracy.

The method is not suitable for semi-dry or poorly compacted concrete, although, in the case of air entrainment, Neville[44] suggests that since the voids are discontinuous they will remain air-filled under vacuum and will absorb no water. If this is the case, entrained air will not affect the results, which are influenced only by capillary voids. Where the aggregates are very porous or contain an appreciable amount of combined water, however, the corrections required will be so large that the results may be of little value. This shortcoming is likely to limit the value of the method for artificial aggregates.

2.9.5 Air Content

This is normally determined by microscopic study of a polished dye-impregnated cut surface of concrete, using procedures defined by ASTM C457-80,[25] as discussed previously.

2.10 PERMEABILITY

Permeability of the interior of a body of concrete can be measured only by means of laboratory flow tests upon a sample of concrete. Several techniques which permit an assessment of the permeability of concrete in the surface zone to water, air, carbon dioxide or other gases under pressure are, however, available or under development. They provide practical ways of assessing the permeability of surface zone concrete under *in-situ* conditions, and it is now recognised that information of this type may be particularly valuable as an indicator of potential durability.

The most widely established method is a non-destructive test relating to surface zone water penetration rate. Called the initial surface absorption test, it is described in BS 1881 Part 5[7] and discussed in detail elsewhere.[14] Initial surface absorption involves measurement of the rate of flow of water per unit area into a concrete surface subjected to a constant applied head. The equipment consists of a cap which is clamped and sealed to the concrete surface, with an inlet connected to a reservoir and an outlet connected to a horizontal calibrated capillary tube and scale, as illustrated in Fig. 2.15. Measurements are made of the movement of the water in the capillary tube over a fixed period of time following closure of a tap between the cap and the reservoir.

The absorption of water by a dry surface is initially high but decreases as the water-filled lengths of capillaries increase; thus measurements have to be taken at specified time intervals from the start of the test.

It is essential to provide a watertight seal between the cap and the concrete surface, and difficulties are likely to be encountered. Sometimes it will be necessary to drill the surface for fixings. Results are affected by variations in moisture content of the concrete, and problems of standardising *in-situ* moisture conditions may limit site measurements to comparative situations, although the Concrete Society has recently provisionally suggested durability classifications based on measured values obtained by this method.[45]

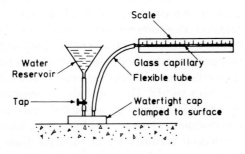

Fig. 2.15.   Initial surface absorption test.

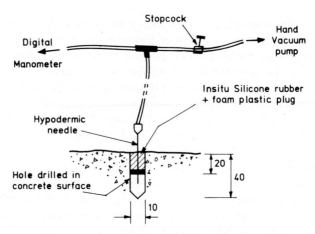

Fig. 2.16.   Modified Figg air permeability test.

Some damage will be caused by the other available methods,[45] which all require a hole to be drilled into the concrete surface. The most widely known in the UK is the modified Figg[46] air permeability method, illustrated in Fig. 2.16. This is based on the time taken for a pressure change from 55 to 50 kN/m² below atmospheric to occur within the evacuated sealed void. Although the use of tests of this type is increasing, experience in the interpretation of site results is still limited.

2.11   PERFORMANCE TESTING, MONITORING AND ASSESSMENT OF REPAIRS

Dynamic response testing of entire structures is commercially available and may be used to monitor stiffness changes due to cracking and deterioration. The 'dynamic signature' of a structure may be obtained by monitoring the response at critical points to a known imposed impulse or vibration applied to another part of the structure. This may be particularly useful to monitor the effects of a suspected overload or deterioration with time.

A wide range of surface measurement techniques are available to assess surface strains and crack widths. These include mechanical, acoustic and electrical resistance gauges and optical equipment which may be utilised either to monitor performance during a specific loading sequence, intended to demonstrate structural performance or resulting from an abnormal load, or to monitor long-term behaviour. The use of acoustic emission measurements to monitor internal crack development may be worthwhile in such cases as a warning of impending failure. However, the Kaiser effect is not reliable for concrete, because of its partial self-healing properties, and stress levels or previous stress histories cannot be determined, thus limiting the value of this approach. Geometric changes associated with overall structural performance may be monitored by both optical and electronic techniques, which may be automated for long-term surveillance.

The concrete repair industry has expanded rapidly in the past few years, and the need to develop standard testing procedures to permit routine testing of repairs is now recognised. Testing of the quality of the cured repair material in-place is necessary, as well as assessment of the integrity of bond between the base material and the repair. Clearly it should be possible to develop pull-off,[31] pulse velocity and pulse-echo techniques for the latter purpose, although difficulties may be encountered with small patch repairs. Dynamic response testing of entire structures or structural components before and after repairs has also been proposed as a technique for assessing the effect of major repairs upon structural stiffness, and is commercially available. Development of *in-situ* non-destructive tests of the quality and properties of the repair material itself is more complex and research is in progress in this area. Little information is currently available upon the interactions between repair medium and base material which are necessary to prevent further deterioration, especially where repairs have been required as a result of reinforcement corrosion, and this is a further area of current research.

2.12   GENERAL OBSERVATIONS AND FUTURE DEVELOPMENTS

Despite a significant increase in the need to test concrete bridges in recent years, due to a large extent to problems of durability, some aspects of development of commercially available testing equipment and procedures have been surprisingly slow. Equipment for measuring the cover and location of reinforcement, for example, has changed little despite considerable scope for improvements. The need for reliable tests to identify hidden internal features of a concrete member when only one face is accessible (e.g. voids, cracks, delamination etc.) is well established and, although research is in progress to develop the application of techniques such as short-wave radar, pulse-echo and reflected-wave ultrasonics, progress is slow and hampered by lack of funding.

Some UK research activity is being directed towards test methods concerned with the presence and likelihood of reinforcement corrosion, including measurement of resistivity, half-cell potential, gas and water permeability, and chloride ion diffusion. Some of these techniques are now used on site, but only limited concerted effort has been directed towards consolidation and dissemination of experience. Research into the *in-situ* assessment of alkali–silica reaction has also been hampered by lack of funding in the UK, although some progress has been made in other countries, including Denmark.

Whilst recent developments of equipment have been few, developments in technique and applications for surface hardness, ultrasonic pulse velocity and radiographic measurements are at present being incorporated into revisions of the relevant British Standards (BS 1881 Parts 202,[8] 203[9] and 205,[11] respectively). Developments in chemical analysis are being incorporated into revisions of BS 1881 Part 6, which will be republished as BS 1881 Part 124 in due course.

The one area in which there have been several new tests developed is that of near-to-surface strength assessment. These have generally been variations of long-established concepts aimed at some specific application, and such developments seem likely to continue, although the scope for significant improvements in accuracy of strength estimation would appear to be limited.

There is little doubt that *in-situ* testing of concrete may offer much valuable information to engineers concerned with the construction and maintenance of bridges. The necessity of careful planning to achieve the greatest benefits has been emphasised in Chapter 1, but it is also important to recognise the limitations to that which can be achieved and not to fall into the trap of expecting the impossible. Given this awareness, it seems likely that the use of *in-situ* testing will continue to develop in its application to concrete bridges for the foreseeable future.

REFERENCES

1. Woodward, R. J., Inspecting concrete bridges, *J. Inst. Phys.*, 149–51 (1984).
2. Turton, C. D. *et al.*, *Non-structural Cracks in Concrete*, Tech. Rep. 22, Concrete Society, London (1982).
3. Pollock, D. J., Kay, E. A. and Fookes, P. G., Crack mapping for investigation of Middle East concrete, *Concrete*, 15(5), 12–18 (1981).
4. Higgins, D. D., Diagnosing the causes of defects or deterioration in concrete structures, CP sheet 69, *Concrete*, 15(10), 33–4 (1981).

5. British Standards Institution, BS 1881 Part 201, *Testing concrete—Guide to the use of non-destructive methods of test for hardened concrete*, London (1986).
6. British Standards Institution, BS 4408, Parts 1–5, *Non-destructive methods of test for concrete*, London (1969–74).
7. British Standards Institution, BS 1881 Part 5, *Methods of testing concrete for other than strength*, London (1970).
8. British Standards Institution, BS 1881 Part 202, *Testing concrete—Recommendations for surface hardness testing by rebound hammer*, London (1986).
9. British Standards Institution, BS 1881 Part 203, *Testing concrete—Recommendations on measurement of the velocity of ultrasonic pulses in concrete*, London (1986).
10. British Standards Institution, BS 1881 Part 204, *Testing concrete—Recommendations on the use of electromagnetic cover measuring devices*, London (1986).
11. British Standards Institution, BS 1881 Part 205, *Testing concrete—Recommendations for radiography of concrete*, London (1986).
12. British Standards Institution, BS 1881 Part 206, *Testing concrete—Recommendations on the determination of strain in concrete*, London (1987).
13. British Standards Institution, BS 1881 Part 208, *Testing concrete—A test for determining the initial surface absorption of concrete*, London (1987).
14. Bungey, J. H., *The Testing of Concrete in Structures*, Surrey University Press, Glasgow (1982).
15. British Standards Institution, BS 1881 Part 120, *Testing concrete—Method for determination of the compressive strength of concrete cores*, London (1983).
16. Ladner, M., *Insitu load testing of concrete bridges in Switzerland*, SP 88-4, American Concrete Institute, Detroit, 59–80 (1985).
17. Ladner, M., *Unusual methods for deflection measurements*, SP 88-8, American Concrete Institute, Detroit, 165–80 (1985).
18. British Standards Institution, BS 1881 Part 6, *Analysis of hardened concrete*, London (1971).
19. Power, T. O. and Hammersley, G. P., Practical concrete petrography, *Concrete*, **12**(8), 27–30 (1978).
20. Anon., Concrete problems burst forth, *New Civil Engineer*, 24 April 1980.
21. Wood, J. G. M., Engineering assessment of structures with alkali–silica reaction, *Proc. Conf. on Alkali–Silica Reaction*, Concrete Society, London (1985).
22. Hawkins et al., *Minimising the risk of alkali–silica reaction*, Draft Concrete Society Technical Report, Concrete Society, London (1985).
23. American Society for Testing and Materials, ASTM C227-81, *Test for potential alkali reactivity of cement–aggregate combinations (mortar bar method)*, Philadelphia (1981).
24. American Society for Testing and Materials, ASTM C289-81, *Standard test method for potential reactivity of aggregates (chemical methods)*, Philadelphia (1981).
25. American Society for Testing and Materials, ASTM C457-80, *Air void content in hardened concrete*, Philadelphia (1980).

26. Concrete Society, *Assessment of fire damaged concrete structures and repair by gunite*, Tech. Rep. 15, London (1978).
27. British Standards Institution, BS 8110 Part 2, *The structural use of concrete, Code of Practice for use in special circumstances*, London (1985).
28. Vassie, P. R., *A survey of site tests for the assessment of corrosion in reinforced concrete*, Lab. Rep. 953, TRRL, Crowthorne (1980).
29. Vassie, P. R., Reinforcement corrosion and durability of concrete bridges, *Proc. ICE*, Part 1, **76**, 713–23 (1984).
30. Figg, J. W. and Marsden, A. F., *Development of inspection techniques of reinforced concrete—a state of the art review of electrical potential and resistivity measurements for use above the water level*, HMSO, OTH-84-2405 (1985).
31. Petersen, C. G. and Poulsen, E., Insitu test methods for concrete with particular reference to strength, permeability, chloride content and disintegration, *Proc. Int. Conf. deterioration and repair of reinforced concrete in the Arabian Gulf*, Bahrain Soc. of Engineers/CIRIA, 495–508 (1985).
32. Roberts, M. H., *Carbonation of concrete made with dense natural aggregates*, Building Research Establishment, IP 6/81 (1981).
33. Somerville, G., The design life of concrete structures, *Struct. Engineer*, **64A**(2), 60–71 (1986).
34. Kunz, J. T. and Eades, J. W., *Remote sensing techniques applied to bridge deck evaluation*, SP 88-12, American Concrete Institute, Detroit, 237–58 (1985).
35. British Standards Institution, BS 5400 Part 4, *Code of Practice for design of concrete bridges*, London (1984).
36. Bungey, J. H., *The influence of reinforcement on ultrasonic pulse velocity testing*, SP 82-12, American Concrete Institute, Detroit, 229–46 (1984).
37. Bungey, J. H., Determining concrete strength by using small diameter cores, *Mag. Conc. Res.*, **31**(107), 91–8 (1979).
38. British Standards Institution, BS 6089, *Guide to assessment of concrete strength in existing structures*, London (1981).
39. Concrete Society, *Concrete core testing for strength*, Tech. Rep. 11, London (1976).
40. Domone, P. L. and Castro, P. F., An expandable sleeve test for insitu concrete strength evaluation, *Concrete*, **20**(3), 24–5 (1986).
41. Chabowski, A. J. and Bryden-Smith, D. W., Assessing the strength of insitu Portland cement concrete by internal fracture tests, *Mag. Conc. Res.*, **32**(112), 164–72 (1980).
42. Long, A. E. and Murray, A. McC., *The pull-off partially destructive test for concrete*, SP-82, American Concrete Institute, 327–50 (1984).
43. Lees, T., *Analysis of hardened concrete*, Cement and Concrete Association, Slough, TDH 8477 (1981).
44. Neville, A. M., *Properties of Concrete*, Pitman, 152 (1977).
45. Concrete Society, *Insitu permeability of concrete—a review of testing and experience*, draft report, London (1985).
46. Cather, R., Figg, J. W., Marsden, A. F. and O'Brien, T. P., Improvements to the Figg method of determining the air permeability of concrete, *Mag. Conc. Res.*, **36**(129), 241–5 (1984).

3

Assessment of Load Effects in Reinforced Concrete Slab Bridges

R. J. COPE

Department of Civil Engineering, Plymouth Polytechnic, UK

3.1 INTRODUCTION

Many small and medium span bridges have been built as reinforced concrete slabs. Assessment of their likely structural response to loading requires a different procedure to that adopted in their design. When assessing the response of an existing slab bridge, the structural form is already determined. It will have been subjected to an unknown load history, including the effects of environmental changes, possible movement of supports and the effects of the construction process, in addition to the effects of gravity loading. An assessment may also be complicated by defects in design, detailing or construction, and by chemically- or physically-induced deterioration. A visual inspection and assessment is, therefore, a necessary preliminary to any structural assessment.

Design of a slab bridge is usually performed using contemporary specified loading, contemporary analytical techniques, and contemporary specified procedures for checking strengths and for detailing reinforcement to ensure adequate ductility. With increasing knowledge and changing requirements, the specifications change. If ductility is reduced due to inadequate detailing standards or the effects of deterioration, simplified strength checks and analytical procedures can be invalidated.

Assessing a structure may, therefore, provide an engineer with the increasingly rare opportunity to use 'first principles', and it may be necessary to determine the reasoning behind a code formula in order to extrapolate it to the prevailing conditions in the particular structure under examination.

Assessments of structural performance and of the safety factor against

63

collapse require an appraisal of the loading, prediction of the effects of the loading on the structure and the calculation of the structural capacity. To determine overall safety factors, using limit state procedures, partial safety factors are required to take into account uncertainties: in loading intensities; in the accuracy of analytical predictions; and in the knowledge of material properties.

In the USA, the term limit state is not used in design codes. However, service load design and strength design are considered separately, and these requirements may be considered to be equivalent to the serviceability and ultimate limit states.

In general, consideration of ultimate limit state criteria will be of most importance. Because of the inherent ductility of a concrete slab bridge, the upper and lower bound methods of limit analysis can be used to assess the flexural capacity. The assessment of shear capacity in the vicinity of bearings and columns can be more difficult, however, as the existing load acting on any one bearing may be difficult to predict accurately and a slab has only a limited capacity for redistribution of load between bearings.

Quantities required for assessing serviceability conditions, such as deflections and stress levels in concrete and reinforcement, are difficult to determine accurately. However, existing crack widths can be measured and changes in deflections and crack widths can be monitored. Probably the most important serviceability consideration for a slab bridge is that to determine whether deflections are increasing because stress levels are producing plastic strains in the reinforcement.

The widespread use of computer-based methods of analysis has greatly increased an engineer's analytical capability. However, the reliability of analytical predictions is dependent on proper account being taken of material properties and support stiffnesses. The recent development of non-linear analytical methods for concrete slabs, which can accommodate changing material stiffness properties with increasing load intensity and which can follow the development of cracking in concrete and yielding of reinforcement, has greatly increased the potential for a realistic prediction of load distribution. However, because of the relatively high costs of such methods, it is likely, at least for the present, that they will be reserved for critical cases.

3.2   PARTIAL SAFETY FACTORS

Strictly put, the aim of an assessment of a slab bridge is to determine whether there is an acceptable probability that it will not become unfit for its specified use during its specified life. To achieve this, it is necessary to assess the response of the bridge to working loads and to determine the factor of safety against failure. To be able to determine an acceptable probability, loading, structural response and material properties must be treated as variable properties. A comparative review of partial safety factors used in British, European and North American structural design codes has been provided by CIRIA.[1]

For limit state design to British codes, it is usual to define loads in terms of their characteristic values, which are defined as those with a 5% chance of being exceeded during the 120-year design life of a bridge. However, for bridges, there are insufficient statistical data to determine characteristic values for all loads, and so codes of practice define nominal loads Q_k to take their place. Design loads Q^*, at each limit state, are obtained by multiplying the nominal loads by a partial safety factor γ_{fL}. Thus, $Q^* = Q_k \gamma_{fL}$.

Design load effects S^* (moments, shear forces etc.), which must be resisted at a particular limit state, are obtained by multiplying the calculated effects of the design loads by a partial safety factor γ_{f_3}. Thus, $S^* = \gamma_{f_3}$ (effects of $Q_k \gamma_{fL}$).

The characteristic strength of a material, f_k, is defined as that strength with a 95% chance of being exceeded. For reinforcement, this definition is clear, but it is less so for concrete. This is because the characteristic strength of the concrete is based on the results of standard specimen tests. If, for the purposes of an assessment, the characteristic *in-situ* concrete strength is determined, an adjustment is necessary to determine the corresponding characteristic strength of a standard specimen. Details of the procedure can be found in Chapter 2.

Design strengths R^* are, generally, functions of the characteristic standard specimen strengths divided by partial safety factors γ_m. Thus, $R^* = \text{function } (f_k/\gamma_m)$. For example, BS 5400: Part 4[2] gives the design bending resistance of a singly reinforced section to be the lesser of

$$R^* = M_u = \left(\frac{f_y}{\gamma_{ms}}\right) A_s \left[1 - \alpha_1 \frac{(f_y/\gamma_{ms}) A_s}{(f_{cu}/\gamma_{mc}) bd}\right] d \tag{3.1a}$$

and

$$R^* = M_u = \alpha_2 \left(\frac{f_{cu}}{\gamma_{mc}}\right) bd^2 \tag{3.1b}$$

where $f_y, f_{cu} = f_k$ for the steel and concrete respectively, and $\gamma_{ms}, \gamma_{mc} = \gamma_m$ for the steel and concrete respectively; α_1 and α_2 are concrete stress block parameters.

However, in some cases, the design strength is obtained by dividing the resistance of the reinforced concrete section by γ_m. Thus, $R^* =$ function $(f_k)/\gamma_m$. For example, the design shear resistance of a member with no shear reinforcement is given by BS 5400[2] to be

$$R^* = V_c = 0.27\left(\frac{100A_s}{bd}\right)^{1/3}(f_{cu})^{1/3}bd/\gamma_m \qquad (3.2)$$

However, no matter which way R^* is defined, for satisfactory behaviour

$$R^*(f_k, \gamma_m) \geq S^*(Q_k, \gamma_{fL}, \gamma_{f3}) \qquad (3.3)$$

In the USA,[3] nominal H and HS lane loadings are specified to act on nominal lanes, and load factors are prescribed. Nominal section strengths are factored by strength reduction factors, which depend on the type of failure under consideration. This is similar to the British treatment of shear failure, but is different in principle to the treatment of flexural failure.

When a structure has been constructed, some of the uncertainties regarding the strengths of the materials and the geometry are removed. Information can be obtained from records, from measurements, and from methods of non-destructive and partially destructive testing (see Chapter 2). Also, more accurate assessment of dead loads can be made from measurements on the structure, and the range of live loading may be limited by traffic controls. With reductions in the degrees of uncertainty, it seems reasonable to consider the use of lower values of partial safety factors for an assessment than those used for design.

3.3   LOADING

3.3.1   Dead Loading

To assess the serviceability limit states, BS 5400 permits a unit partial factor of safety γ_{fL} for dead loads. To assess the ultimate limit states, a value of 1·15 is specified for concrete when the dead load is accurately assessed, otherwise a factor of 1·20 is specified. Corresponding values for surfacing dead load are 1·20 and 1·75 respectively. However, with the approval of the authority that is to be responsible for ensuring that the nominal superimposed dead load is not exceeded during the life of the bridge, these values may be reduced to 1·0 and 1·2 respectively.

When assessing a concrete bridge structure, the actual dimensions and material densities can be determined. Ideally, cylinders of concrete need to be obtained from the top and soffit zones to take into account the density variation over the thickness. However, this refinement may not be possible, due to the presence of reinforcement.

With measured data, it seems not unreasonable for an appraisal of the serviceability limit states to use unit values of the partial safety factors with permanent loads. For the ultimate limit states, a value of 1·05 could be used for the dead load and 1·2 for the superimposed dead load. Reducing the dead load partial factor of safety to 1·05 for the ultimate limit state is in line with a recent recommendation for building structures.[4] However, the recent code for the assessment of highway bridges and structures[5] specifies the factor 1·15 for the dead load of concrete bridges in the UK.

3.3.2 Traffic Loading

The nominal live (highway) loadings of BS 5400: Part 2,[6] their associated partial safety factors and the serviceability stress limits of BS 5400: Part 4[2] were calibrated to ensure that designs in the UK would not change substantially from previous practice. HA loading is based on a report by Henderson[7] and is intended to represent possible, normal, actual vehicle loading.[8] In the discussions to Henderson's paper, engineers argued that lower load intensities could be used for the design of slab bridges in less developed parts of the world. However, in view of the subsequent rapid spread of traffic and increasing use of heavier lorries, it would be prudent to consider full HA loading for all slab bridge assessments, unless local data on vehicle loading and loading trends are available. Flint and Edwards[9] suggested that both the normal HA and abnormal HB loadings could be treated as characteristic loads for bridges with a design life of 120 years, for the traffic conditions of the late 1960s.

Flint et al.[10] have reviewed more recent data on vehicle loading in the slow lane of a motorway. For spans of up to 30 m, they report characteristic load events to give moments only about 80% of those obtained using the current HA loading. Armitage[11] reviewed the load effects of modern lorries and made recommendations concerning future vehicles. Although heavier lorries are to be expected, controls on axle spacing are such that their load effects on a bridge lane are not expected to exceed greatly those of the standard loadings.

For the assessment of a bridge in the UK, the Department of Transport[5] has recommended the use of a loading curve which gives an increasing load intensity for loaded lengths of less than 50 m. This distributed loading, which is also a function of the notional lane width, is to be used in conjunction with the knife edge loading specified as part of HA loading.

The loading curve is based on studies of the maximum possible bending moment and shear force effects that could be produced by considering the effects of the full range of vehicles allowed under the Motor Vehicles Construction and Use (C&U) Regulations.[12]

When the specified loading for an assessment is used, abnormal HB loading need not be considered. However, the Standard[5] is mainly intended for the assessment of highway bridges built prior to 1922 and it states that 'When loadings other than those specified in this Standard are considered necessary for assessment purposes, the loadings shall comply with the requirements given in BS 5400: Part 2, as implemented by the Departmental Standard BD 14/82'.[13] It is, therefore, probable that the vast majority of slab bridges will have to be assessed for the effects of abnormal loadings.

At present, there is insufficient data on which to base characteristic loadings involving traffic in more than one lane, and it is likely that the maximum loading under accident conditions will have to be considered for assessment purposes. It is understood that, for UK bridges, the Department of Transport is likely to amend the HA loading specification to allow for the large increase in the numbers of heavy goods vehicles, and for possible lateral bunching of vehicles in relatively wide lanes. The nominal HA loading to be specified is intended to represent the maximum possible loading divided by a partial safety factor of 1·5.

The HB loading is intended to represent abnormally heavy vehicles, which are subject to statutory control, and this type of loading consequently attracts smaller partial safety factors. National regulations specify design vehicles appropriate to local conditions. For the UK, the intensity of the loading varies with the classification of the road being carried. For motorways and trunk roads, 45 units are specified. The number of units for other routes are: 37·5 for principal routes; 30 for other public roads; and 25 for accommodation roads and byways. Church and Clark[14] have estimated, on the basis of data of recorded vehicle movements in the UK during 1979–80, that a typical UK concrete bridge can expect to be crossed by an abnormal vehicle once every three years. This justifies consideration of the effects of abnormal loading when assessing most slab bridges.

3.3.3 Temperature Loading
The load effects of restrained temperature movements have to be considered in combination with highway loading. This is not straightforward, as the forces generated by restrained temperature movements are dependent on the tangent moduli of the materials at the actual stress levels

prevailing. If the temperature-induced bending moments and shearing forces are determined on the basis of linear–elastic structural response, the partial safety factors γ_{fL} can be applied directly to the temperature change. If, however, non-linear or plastic material properties are used, one has to decide whether the temperature change or the restraining forces is the nominal 'loading', in order to apply the loading partial safety factors.

BS 5400: Part 2 specifies different partial safety factors to be applied to the temperature range and to the temperature difference over depth. The former would cause an unrestrained slab deck to expand or contract by equal amounts in all plan directions. Integral columns and piers resist this movement, and in so doing introduce secondary moments and shear forces into the slab deck. The specified partial safety factor is 1·3 for the ultimate limit state, and this relatively high value is applied to temperature changes that must be based on an assessment of the temperature when the restraints first became effective.

Design temperature profiles over depth have been published for a number of countries. Differences in temperature over depth cause curvatures of sections and in a statically indeterminate structure, such as a slab bridge, these lead to additional secondary moments and shear forces. For a non-linear temperature distribution over depth, self-equilibrating stresses can also be set up on each cross-section. However, if it is assumed that there is local plasticity at the critical sections, these can be ignored at the ultimate limit state. BS 5400: Part 2 specifies a unit partial safety factor for use with its prescribed temperature profiles over depth at the ultimate limit state.

3.3.4 Load Combinations

Church and Clark[14] have considered the loading partial safety factors for use when highway loadings are combined with dead loading and the effects of temperature changes. The factors applied to the highway loadings are reduced when the effects of all three loadings are combined, to allow for the reduced probability that a number of loads acting together will all attain their nominal values simultaneously.

The authors reported that the frequency of abnormal loading experienced by a particular bridge depends greatly on its location. For an 'average' trunk road bridge anywhere on the UK mainland, it seems unlikely that severe thermal and force loadings would ever exist simultaneously. However, for the routes most heavily used by abnormal loads, the results indicated that it is very likely that the load combination would occur within the 120-year design life of a bridge.

On the basis of their study, Church and Clark recommend that HB loading need not be considered in combination with temperature loading. They also recommended that the partial safety factors on HA loading be reduced to 0·5 and 0·60 for the serviceability and ultimate limit states respectively. However, in view of the serious consequences of a shear failure of a fully laden slab bridge on a hot summer's day, and the recent data on lane loadings reported by Flint et al.,[10] the writer would suggest that a factor of 0·8 be used to factor HA loading for shear assessment purposes.

3.4   STRUCTURAL ANALYSIS

Linear, non-linear or full range and plastic methods of analysis can all contribute to an assessment of a slab bridge. However, it should be emphasised that the predictions of linear and non-linear methods of analysis must be regarded as approximate. Although these methods can provide enormous quantities of detailed data, their inherent assumptions regarding the structural response are based on the behaviour of idealised materials, idealised boundary conditions and a known construction sequence. Movement of supports, a failed bearing, the effects of cracking during construction and under environmental loading, for example, can all lead to considerable stress redistribution.

Non-linear methods of analysis attempt to predict the response of a slab bridge after cracking, the effects of sustained and repeatedly applied loads, and local yielding of reinforcement. At present, the methodology only deals with changes to the in-plane material properties, and the stiffness to transverse shear is kept constant. It is, therefore, not possible to predict such phenomena as the redistribution of reactions between columns or bearings after the formation of local shear cracking.

Yield line methods can predict the ultimate load of a slab bridge failing in flexure by considering likely failure mechanisms. As only ultimate conditions are considered, the previous stress history is not relevant to the analysis. It is assumed that a slab possesses sufficient ductility for the required stress redistribution to occur, and that a shear failure does not intervene before the flexural mechanism has formed.

When flexural failure of a continuous structure is being assessed, Clark[8] suggests that thermal continuity or secondary moments can be ignored, provided the slab sections possess sufficient ductility.

For bridge design, an unusually low ratio of partial safety factors for loading at the ultimate and serviceability limit states is considered. This has the effect of making serviceability criteria more critical in design. As only ultimate limit state criteria have to be checked in the assessment of an existing slab bridge,[5] assessments of flexural adequacy based on inelastic analysis (including the yield line method) can be advantageous.

Predicted shear forces, including those induced by restraint of thermal movements, are, however, usually based on the results of linear–elastic analyses. Essentially, it is assumed that the shear forces predicted by the relatively peaky moment distributions predicted by linear analyses will be greater than those that occur after moment redistribution. Whilst this assumption appears to lead to satisfactory designs, it may be over-conservative for assessment purposes. However, there is at present no reliable alternative approach. Because of the limited capacity of a slab to redistribute shear forces after a local punching failure at a bearing or column, such a failure is generally assumed to indicate overall structural failure.

3.4.1 Linear Methods of Analysis

Linear methods of analysis can provide reasonable estimates of moment and shear force distributions under working loads. Predictions of linear analyses using loads factored for the ultimate limit state are often used in design. The most reliable linear analytical techniques for slab bridges are the grillage analogy and the finite element method. Both of these approaches can treat plan-forms of any shape, varying depth, and both column and line supports. When suitable stiffness parameters are provided, they can also be used to analyse pseudo-slab bridges of most commonly encountered cross-sections.[15] Both methods are readily available in computer packages and can deal easily with a large number of load cases.

The grillage analogy is usually the cheaper of the two methods. However, for skew and curved slabs, judgement may be required to interpret the results to obtain the equivalent moments and shear forces in the slab bridge whose analysis is required. The grillage model should be based on an orthogonal assemblage of members, wherever possible. Because the method uses equilibrium equations in its formulation, the predicted couples, torques and shear forces acting on the ends of the grillage members automatically satisfy equilibrium conditions.

A finite element solution is based on minimising the total potential energy of a structural system with an assumed, approximate displaced shape. In essence, the method determines values of displacement variables to obtain the minimum energy configuration. Equilibrium of moments and shear forces is not, therefore, guaranteed and should be checked. A sufficiently fine mesh of elements should be selected to provide moments and shear forces that satisfy equilibrium to an acceptable tolerance. Guidance on suitable meshes for particular element formulations is usually provided with standard computer packages. Because bending and twisting moments are predicted directly, no interpretation of results is required for a slab analysis.

A detailed discussion of the structural idealisations and interpretations required for the analysis of slabs by the grillage analogy, and of factors to be taken into account when specifying finite element meshes, can be found in Cope and Clark.[15]

3.4.1.1 Moments

Provided equilibrium equations are satisfied everywhere, and there is sufficient reinforcement at every section to ensure that the resistance in every direction is at least equal to the corresponding applied moment, a bending collapse will not occur. This method of assessment is conservative and is in accordance with the lower bound theorem of limit analysis. This says that a design based on a moment field in equilibrium with the applied loading, and with no violation of the yield criteria at sections, will carry at least that applied loading. However, the method cannot be used to determine the ultimate load that can be supported when there is capacity for moment redistribution after first yielding of reinforcement.

3.4.1.2 Shear Forces

Column forces and moments, and bearing reactions, can be determined from either a finite element or a grillage analysis. To obtain an accurate prediction of moment and shear force predictions, the column and bearing stiffnesses should be included in the analysis. This is especially important for skewed and curved bridge decks, as the distribution of reactions on a line of bearings is strongly influenced by the bearing stiffnesses.

Bearings of an existing bridge should be inspected, if possible, to ensure that they have not deteriorated. If there is any doubt concerning the stiffness of a bearing near an obtuse corner of a slab bridge, it would be prudent to analyse the deck with the stiffness of the suspect bearing set to its maximum likely value.[16]

Analysis of an analogous grillage predicts two stress resultants that produce vertical shear stresses: shear forces and torques. When assessing the flexural–shear capacity, the shear stresses from both of these components need to be taken into account. The critical sections are likely to be near free edges, where the vertical component of the torsional shear stress from one grillage member is not reduced by that from an adjacent, parallel member. To assess the factor of safety against a flexural–shear failure, the shear force per unit length is required. In order to provide a realistic estimate of the shear force distribution on sections close to a free edge, it may be necessary to use relatively closely spaced grillage members parallel to the edge.[17] It is suggested that the outermost two members be spaced a distance apart equal to the slab depth.

The distribution of shear forces predicted near free edges by a finite element solution will depend on the plate theory used in the finite element formulation. If classical, or thin plate, theory is used, the predicted shear forces on a section across a slab are unlikely to satisfy equilibrium. This is because the plate theory does not exactly satisfy the boundary conditions at a free edge.

To ensure that equilibrium is satisfied, and to obtain the correct distribution of shear forces on sections intersecting a free edge, the twisting couples predicted to act on the free edge must be replaced by suitably distanced equal and opposite vertical forces. To obtain a satisfactory solution, a narrow strip of elements should be used along a free edge.

The required shear force acting on a section intersecting a free edge can be obtained as follows. Using the notation of Fig. 3.1, it is assumed that the bending moment per unit length M_t and the twisting moment per unit

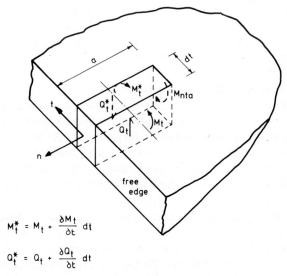

$$M_t^* = M_t + \frac{\partial M_t}{\partial t}\, dt$$

$$Q_t^* = Q_t + \frac{\partial Q_t}{\partial t}\, dt$$

Fig. 3.1.   Stress resultants near a free edge.

length M_{nta} are predicted correctly. The shear force per unit length Q_t is
then calculated to ensure rotational equilibrium of the small block of slab
about the normal to the free edge, with no forces acting on the free edge
(thin plate theory predicts values for both the shear force Q_n and the
twisting moment M_{nt} on the free edge, such that $Q_n + \partial M_{nt}/\partial t = 0$).[15]
    For rotational equilibrium,

$$\int_0^a Q_t \, da \, dt = \int_0^a \frac{\partial M_t}{\partial t} \, dt \, da + M_{nta} \, dt \tag{3.4}$$

Dividing eqn (3.4) by dt and integrating, the total shear force $\bar{Q}_t$, acting on
the section of width a, is given approximately by

$$\bar{Q}_t = \left(\overline{\frac{\Delta M_t}{\Delta t}} \right) + M_{nta} \tag{3.5}$$

where $(\overline{\Delta M_t / \Delta t})$ is the rate of change of the bending moment on sections
normal to the free edge of width a, and can be obtained from the finite
element solution using a first-order finite difference approximation.
    It is suggested that the distance a be set to one slab thickness. Vertical
equilibrium of complete segments with sections across the width of a slab
should be checked to ensure that a sufficiently fine finite element mesh has
been selected to provide acceptable shear force predictions.
    If the finite element formulation is based on Mindlin's plate theory, then,
with a sufficiently fine mesh, the predicted shear forces on a section across a
slab will satisfy equilibrium conditions, and no modifications to the
predicted stress resultants are required.[17]

3.4.2 Non-linear or Full Range Analysis
Although linear elastic solutions can provide moment and shear force fields
for strength assessments, they give only a very approximate description of
slab behaviour. At working load levels, this is due mainly to the variation of
stiffness due to cracking and, when there is a relatively stiff surrounding
structural system, the coupling of membrane and flexural effects. At higher
load levels, there is further considerable redistribution of stresses due to the
spread of yielding in the reinforcement.
    To follow the structural response from initial cracking of the concrete to
overall structural failure, it is necessary to perform a non-linear analysis.
The finite element method provides the natural vehicle for such methods,
and research to produce reliable, if relatively expensive, methods is now
well advanced. These methods could be useful for assessing a slab bridge, as

they can provide an indication of prevailing stresses and of the remaining load capacity.

Attempts have been made to predict the full range response of scale-model skew slab bridges tested under multiple load patterns and with repeated application of loading to the serviceability limit state intensity.[18,19] Although the analyses have been shown to give good predictions of strength and of the failure mode, deflections are predicted less reliably.

In practice, a slab is subjected to an unknown load history; suffers cracking due to changes in environmental conditions; experiences creep under a sustained load component; undergoes growth of cracking due to repeatedly applied loads; and is constructed of a material whose properties vary in a largely indeterminate manner. Attempts to predict a detailed response to loading are, therefore, likely to prove futile. However, if the existing cracked state of a slab can be analytically modelled approximately, there is a reasonable chance that its subsequent response to a single load pattern can be predicted.

Cope and Rao[20] have suggested using a two-stage analysis. The first stage is concerned with predicting the existing strains and the extent of cracking. A non-linear analysis is performed using an estimated long-term modulus of elasticity for the concrete. Permanent loading is applied as a single load increment, and iterations are performed to permit the development of any cracking in the analytical model. After equilibrium has been established, a uniformly distributed loading to represent the maximum likely traffic conditions is applied and a new equilibrium configuration established. This loading is then removed to enable the in-service state of the structure under permanent loading to be determined.

The effects of a particular live load pattern are determined in the second stage of the analysis. Short-term concrete properties are used, with predicted stresses and strains combined with those prevailing at the end of the first stage of the analysis. The intensity of the live loading is increased in steps, with iterations being performed to establish equilibrium structural configurations, until overall structural failure occurs.

The method has been shown to give reasonable predictions for laboratory models.[20] It has also been used to help Merseyside County Council engineers assess a slab bridge that was causing public unease due to large deflections and extensive cracking. The reinforced concrete bridge, which is shown in Fig. 3.2, had a skew of approximately 45° and a skew span of about 120 ft. The reinforcement directions were predominantly parallel and perpendicular to the free edges. The bridge was designed in the early 1960s.

Fig. 3.2.   45°, voided reinforced concrete slab bridge with temporary central support.

Fig. 3.3.   Top surface cracking in an obtuse corner (courtesy of Merseyside County Council).

The intensity of traffic loading used for the first stage of the analysis was determined to give an analytical central free edge deflection of the same order as that experienced by the bridge. The analysis predicted extensive top cracking in the vicinity of the obtuse corners. Exposure of the black top and the waterproof membrane in one of the obtuse corners of the bridge revealed the presence of this cracking (Fig. 3.3). Following an overnight downpour, rain water was found to be seeping from the soffit cracks intersecting the uncovered top cracks.

For the second stage of the analysis, the effects of live loading combinations, factored to the ultimate limit state intensities specified in BS 5400: Part 2,[6] were considered. Selected load patterns were then applied with incremental increases in the load factor, to study the structural response up to failure.

By predicting the likely extent of top cracking, the closeness of the strains in the steel to those necessary to cause yielding, and the ultimate load, the full range analysis provided information to the engineer not readily available by any other means. As a result of this study, and of other investigations, the bridge was strengthened by the addition of a central pier. Jacks were inserted to enable the pier to take a part of the dead loading.

3.4.3 Yield Line Analysis

Yield line theory belongs to the upper bound methods of limit analysis. That is to say, it predicts estimates of the collapse loading that are either equal to or greater than the plastic theory collapse load. Details of the method can be found in Cope and Clark,[15] or in specialist texts. Detailed experimental tests have shown that the method can provide reasonable estimates of the ultimate load for slabs that possess sufficient ductility and which do not fail prematurely in shear.

The first stage of a yield line analysis is to propose a valid collapse mechanism. This consists of lines, along which all of the tension reinforcement is assumed to be yielding, and between which are rigid portions of slab. Typical yield line mechanisms for slab bridges are illustrated in Fig. 3.4; further possibilities are shown in Cope and Clark[15] and in Clark.[8]

A mechanism under investigation is moved through a small, virtual displacement. In practice, a convenient point is given a unit displacement. This enables the deflections of the loading and the rotations of the yield lines to be computed. The work done by the loading is then equated to the dissipation of energy in the yield lines to give the value of the loading to cause collapse.

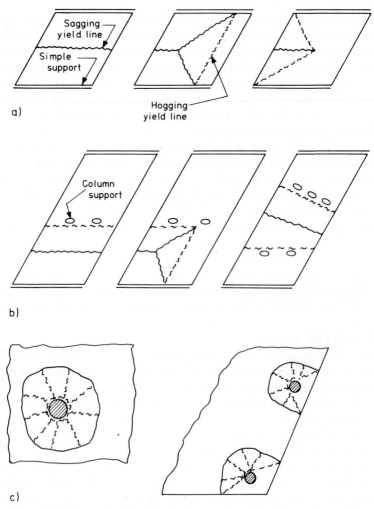

Fig. 3.4.   *Yield line mechanisms: (a) single span skew slabs; (b) continuous slabs; (c) isolated columns.*

If a mechanism requires values of positional parameters defining the locations of junctions of yield lines in a mechanism to be determined first, the loading is expressed as a function of these positional parameters. The minimum value of the collapse loading for the proposed mechanism can then be found by differentiating the function with respect to each positional parameter in turn and equating the resulting expressions to zero. This process results in a set of simultaneous equations which can be solved to give the critical values of the positional parameters, and hence the required value of the loading can be calculated. As the simultaneous equations may be non-linear, an iterative solution may be useful.

It is necessary to consider a number of different possible collapse mechanisms to determine the minimum value of the collapse loading of a slab bridge. A major disadvantage of the yield line method is that an engineer may miss the critical mechanism. However, guidance on a number of likely critical mechanisms has been provided by Clark.[8] A full range analysis does not require selection of the critical mechanism, and will automatically predict the correct solution for a particular load pattern.

Clark has provided a useful bibliography, with worked examples of the yield line method applied to slab bridges. However, when assessing the flexural strength of an existing slab bridge, the values selected for the partial safety factors may need to be modified to take into account some of the points made in this chapter.

3.4.4 Analytical or Structural Performance Factor

The design load effects (for example, moments and shear forces) are obtained by multiplying the analytical predictions by a partial safety factor, γ_{f_3}. This is a factor to take into account inaccurate assessment of the effects of loading, including accidental eccentricities. If a linear relationship is assumed between the loading and a load effect, then the analysis can be conducted using the nominal loads factored by $(\gamma_{fL} \cdot \gamma_{f_3})$. Otherwise, the predicted effects of the design loads are multiplied by γ_{f_3}. For both linear–elastic methods and rigid–plastic yield line methods, a linear relationship may be assumed.

When elastic methods of analysis are used to determine the distribution of forces throughout a structure, BS 5400 specifies a unit value for γ_{f_3} for the serviceability limit states, and a value of 1·10 for the ultimate limit state. However, when plastic methods of analysis such as the yield line method are used, γ_{f_3} should be taken as 1·15. When load effects are determined using the loading specified in BD 21/84, γ_{f_3} is specified to be 1·1.

If a non-linear method of analysis is used, the selection of an appropriate value for γ_{f_3} is at the engineer's discretion. At the ultimate limit state, a lower bound solution can be obtained and the moment distribution is likely to be more realistic than that predicted by a linear analysis. Non-linear methods have definite advantages over the more traditional linear–elastic and rigid–plastic methods. The possibility of specified strain limits being exceeded can be checked. The correct failure mechanism for a particular load pattern is found automatically. Local and global effects are automatically combined, and possible coupling between flexural and in-plane stiffnesses is included. However, the potential of the method is limited by inherent uncertainties in the specification of material properties and the stress state before the application of a particular load increment.

It is suggested that a non-linear analysis of a slab bridge should be carried out using average values for materials properties, rather than characteristic or design values. The moments predicted at the ultimate limit state load intensity would be taken as design load effects (i.e. $\gamma_{f_3} = 1$). The shear forces acting on sections would, however, also be assessed separately, using a linear analysis, with the characteristic values of the material properties, and with a relatively high value of γ_{f_3}, which is discussed later.

For a determinate structure, no matter which analytical method is used, it seems reasonable to use a unit value for γ_{f_3}, since there should be no error in predicted ultimate moments and shear forces. It can also be argued that a unit value should be used when checking the section strengths of a slab bridge using moments predicted by linear analysis, since errors in analysis are covered by the ability of a slab to redistribute moments.

Whereas it might be thought prudent to use the relatively high value of $\gamma_{f_3} = 1\cdot15$ when an overall collapse mechanism of an indeterminate structure is considered, its use is difficult to justify when a local mechanism under a wheel load is investigated. This is especially the case for top slabs of bridge decks with downstand beams or box webs, as membrane enhancement can then provide the slab with substantial reserves of stiffness and strength.

It should be stressed that using γ_{f_3} as a general multiplier for the moments in indeterminate structures, including wide slabs, is not always meaningful. This is especially the case in zones where moments change sign. Increasing a predicted sagging moment, for example, does not provide for a hogging moment capacity that a more accurate analysis might require.

For a flexural appraisal of an indeterminate structure, the use of a non-unit value for γ_{f_3} could be reserved for overall collapse and for failures that could cause loss of life. In these cases, taking a value of $1\cdot15$ would be prudent. This value is consistent with that recently recommended for primary members of building structures.[3]

3.4.4.1 Prediction of Shear Forces

A flexural–shear strength assessment requires knowledge of the greatest value of shear force applied to a section. That could occur either before or after moment redistribution, and a full range analysis is therefore useful.

For skew and curved slabs, the applied shear force intensity can vary considerably across a section.[17] When shear capacities are to be based on consideration of beam strips, the difficulty, on which codes give no guidance, is to determine the section width to consider. Cope[16] has provided some data to assist engineers in determining critical widths for skew slabs, but data are scarce and the exercise of considerable engineering judgement is required. As the capacity of a slab bridge without shear reinforcement to redistribute shear forces is believed to be very small, it would be prudent to use a relatively high value of γ_{f_3} to factor the predicted shear forces in order to reflect uncertainties in the prediction of local shear force concentrations and the consideration of section widths. A value in the range 1·20–1·25 would be prudent.

Factors of this magnitude are already considered for punching shear calculations, but their justification is obscure. In CP 110[21] the design ultimate value of the concrete contribution to the shear resistance at an internal column with no moment transfer was reduced by 20%. It thus appeared that the material partial safety factor was being modified. However, in a revision to CP 110 and in BS 8110: Part 1,[22] the predicted shear force at an internal column with no moment transfer is increased by 25% and the concrete shear strength is restored to its calculated value. This amendment not only led to the possibility of more shear reinforcement being required, but also changed the nature of the partial safety factor. In BS 8110, it can be regarded as an analytical factor to take into account accidental eccentricities and analytical uncertainties. This procedure is more logical.

BS 5400 neither increases the shear force nor does it reduce the calculated concrete shear capacity at an internal bearing. However, it does require the concrete contribution to the punching shear resistance at edge column bearings to be reduced by 20%. This is an illogical approach, as the purpose of the factor must be to take into account accidental eccentricities and analytical uncertainties. It is suggested that the calculated bearing reaction be increased by $\gamma_{f_3} = 1·25$ for assessment purposes and used with the full, calculated concrete shear resistance.

3.5   MATERIAL PROPERTIES AND SECTION CAPACITIES

Material properties are required both for analytical purposes and to determine values of section strengths. Many researchers have proposed uniaxial (and equivalent uniaxial relationships for biaxial stress states) for concrete, and some of these have been reviewed by Cope.[23] The British bridge code[2] provides uniaxial stress–strain curves based on specimen characteristic cube strengths and material partial safety factors for use with linear–elastic and yield line theories.

In European practice, at each limit state, the design strengths of materials are generally obtained by dividing the characteristic strengths of standard specimens by partial safety factors γ_m. The values specified for γ_m are obtained from two subsidiary partial safety factors which cover possible reductions in the strength of the materials in the structure, as a whole, compared with the characteristic values obtained from tests on control specimens and possible weaknesses due to manufacturing tolerances and other unspecified factors. However, in some cases, for example shear strength, the design resistance of a section is specified directly using experimental data. In America, the design strengths of a section for bending and shear are obtained by multiplying nominal section strengths by the appropriate strength reduction factors. Klein and Popovic[24] have suggested that the values of the strength reduction factors specified for design purposes could be amended, in the light of field survey data, to reflect actual rather than possible variations in quality.

3.5.1 Linear Analysis

BS 5400 tabulates secant values of the concrete modulus with characteristic normal weight concrete cube strengths up to $60\,\mathrm{N/mm^2}$, for use with an elastic analysis 'in the absence of special investigations'. BD 21/84 suggests that, for concrete bridges built between 1918 and 1939, the long-term modulus be taken as $14\,\mathrm{kN/mm^2}$. However, for an assessment of an existing slab, the dynamic modulus for the concrete can be determined to an accuracy of about 10% using ultrasonic pulse velocity measurements taken on site (see Chapter 2).

To determine the effects of imposed deformations and to calculate deflections, BS 5400 is more vague, and recommends the use of a modulus intermediate between the values tabulated and half of those values. The same recommendation is made for determining crack widths and stresses at a section, due to the combined effects of permanent and short-term loading and imposed deformations.

For analytical purposes, the static secant modulus based on the characteristic strength of the concrete is recommended for use with a partial safety factor γ_m of unity, for all limit states. This is believed to be because the moment and shear force distributions from a linear–elastic analysis are proportional to the ratios of the orthotropic stiffness properties of a slab, rather than to their absolute values. The use of the characteristic cube strength of the concrete, rather than the average value, when determining the stiffness parameters for the entire slab bridge is not significant, as the modulus is insensitive to relatively small changes in the cube strength. However, particularly for skew slab bridges and continuous slab bridges supported on discrete bearings, the moment and shear force distributions are also dependent on the ratios of the slab stiffnesses to the bearing stiffnesses, and the most accurate estimates of stiffnesses should, therefore, be used. At high stress intensities, the slab stiffnesses reduce relative to those of the bearings, and this causes the distributions of moments and shear forces to be different to those at low stress intensities. The effects of varying stiffness moduli with stress intensity over a slab can only be predicted using a non-linear method of analysis.

3.5.2 Non-linear Analysis

To enable a non-linear analysis to give reasonable predictions, the concrete and steel moduli should be determined from realistic stress–strain curves. If those specified in BS 5400 are used, the partial safety factor γ_m is set to unity. Alternatively, a concrete stress–strain curve suggested in Part 2 of the Building Standard BS 8110 or in Cope[23] could be used.

The above suggestion of setting γ_m to unity for a non-linear or full range analysis is different to that recommended by BS 8110: Part 2[25] for non-linear methods of *analysis* for the ultimate limit state. The Standard recommends that the material strengths at critical sections within a structure (i.e. sections where failure occurs or where hinges develop) be set to their design strengths for the ultimate limit state, while the materials in all other parts of the structure are at their characteristic strengths. The main drawback of this approach, however, is that the critical sections may not be known in advance. Also, setting the stiffnesses of critical sections of an indeterminate structure to relatively low values would tend to divert loading elsewhere and could thus distort the predicted structural response.

BS 8110: Part 2 recognises that there may be problems with its recommendation, and goes on to suggest that 'If it is difficult to implement within the particular analytical method chosen, it will be acceptable, but conservative, to assume that the whole of the structure is at its design strength'. This approach could be used for predicting overall flexural collapse loads using the yield line method or a full range non-linear analysis, but the writer would expect more realistic predictions of moment and shear force distributions to be obtained from a full range analysis using stress–strain relationships related to average material properties of the concrete, reinforcing steel and bearings, and with γ_m set to unity.

3.5.3 Use of Measured Material Properties

Whenever code-based material property values are used for an assessment, rather than measured values, the actual value of the material partial safety factor becomes indeterminate. This is because concrete properties depend on the quality of local aggregates and cements, and not just on the concrete strength. McGowan and McCafferty,[26] for example, state that for Hong Kong concrete the initial tangent modulus is $5\cdot0\sqrt{f_{cu}}/\gamma_m$ rather than $5\cdot5\sqrt{f_{cu}}/\gamma_m$; the secant modulus for short-term loading for grade 45 concrete is 26 kN/mm^2 rather than 32·5 kN/mm^2; the coefficient of thermal expansion is $9 \times 10^{-6}/°C$ rather than the nominal value of $12 \times 10^{-6}/°C$; and the shrinkage strain is typically four times higher than that recommended by BS 5400. For UK bridges, values of the thermal coefficient to be used for reinforced concrete made with aggregates from a natural source have been tabulated against the aggregate type used.[13] These values range from $13 \times 10^{-6}/°C$ for quartzite to $9 \times 10^{-6}/°C$ for limestone.

Using assumed rather than measured coefficients of thermal expansion and shrinkage strain could also result in indeterminate values of the actual partial safety factor for loading due to temperature changes and shrinkage. This could be important for slab bridges supported on integral columns, for example, since the shear forces around the columns can be significantly affected by the secondary thermal moments and shear forces, which are considered to be linear functions of the coefficient of expansion and the elastic modulus.

3.5.4 Flexural Strength

For use in design, and presumably appraisal of the elements to resist the moments calculated by an analysis, BS 5400: Part 4 (1984) specifies stress–strain curves as functions of material characteristic strengths. BD 21/84 states that, for pre-1939 concrete, it may be assumed that f_{cu} is not more than 15 N/mm^2. The curves for both concrete and steel reinforcement are also specified as functions of the material partial safety factors, γ_m.

Bungey[27] has collected data from a variety of sources to show the concrete strength variation trends over the depth of a structure. His findings are summarised in Fig. 3.5, and it can be seen that the specified value of 1·5 for γ_m at the ultimate limit state gives a reasonable average of the strengths for different structural types. However, it can also be seen that there are considerable variations in strength about that average value. Variations of strength in plan tend to be random in nature and to be of a smaller magnitude than those over depth.

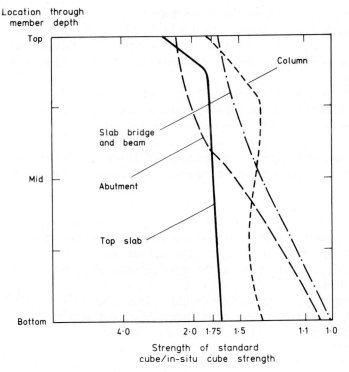

Fig. 3.5.   *Relative concrete strengths over depth.*

The curves shown in Fig. 3.5 suggest that dividing the characteristic cube strength by $\gamma_m = 1.5$ to determine the equivalent *in-situ* characteristic cube strength for use with sagging moment calculations for beams and slab bridges is reasonable. However, for sagging moment calculations of thinner top slabs, a relatively high value of 1.75 for γ_m would not be unreasonable. For hogging moments a value of γ_m of 1.1 could be considered for slab bridges and beams, but for thinner top slabs the value of 1.75 suggested for sagging moments is again appropriate.

However, in certain circumstances, use of these values may be regarded as pessimistic, as tests conducted by Sturman *et al.*[28] have indicated greater ductility and strength for concrete subjected to a strain gradient. They suggest that the improvements in behaviour are due to the confining action of the less highly stressed concrete, which delays the appearance of transverse micro-cracking. For concrete in compression, the presence of an orthogonal compressive stress also leads to stiffer response and to greater strength. This is thought to be due to the effect of Poisson's ratio and, at higher stress levels, to the delay of micro-cracking. For further information on the stiffness and strength of concrete in slabs, the reader is referred to Cope.[23]

The Institution of Structural Engineers[4] recommends that consideration should be given to reducing the γ_m value for concrete to 1·25 for the bending strength appraisal of under-reinforced beam and slab sections in building structures. Thus, there seems to be no objection in principle to varying the γ_m value of 1·5 specified in design codes. However, if lower values were taken, it is suggested that the strength variations over thickness be taken into account, as described above, and also that consideration be given to the nature of the stresses existing at a particular location in a slab.

The value of γ_m of 1·15 specified for steel reinforcement is based on good quality control of production. It is usually impractical to extract sufficient bar samples from an existing structure for testing and it would, therefore, be difficult to justify reducing the specified value of γ_m for the steel at a specified section.

Exposure of a bar surface should indicate whether the steel used was mild, hot rolled or cold worked. Today, the quoted characteristic strengths of reinforcement (f_y) in the UK are 250 N/mm^2 for mild steel and 460 N/mm^2 for high yield steel. This has not always been the case, however. Until very recently, the strength of hot rolled high yield steel (ribbed bars) was quoted at 410 N/mm^2, and that of cold worked high yield steel (square twisted bars) was quoted at 460 N/mm^2 for diameters less than 16 mm and at 425 N/mm^2 for diameters greater than 16 mm. Slab bridges built in the UK before 1939 were usually reinforced with mild steel, whereas those built after 1946 were generally reinforced with a high yield steel. If records indicating the steel characteristic strengths are not available, it would be prudent to use the relevant values quoted by the particular national standard contemporary with the structure.

Since yield line analysis of a slab bridge involves reinforcement over a considerable portion of a structure, it seems reasonable to base the flexural capacity per unit length on higher steel strengths than would be used for an individual section check. Beeby[29] has calculated that, if the value of γ_m for two bars crossing a critical section is 1·15, then, for a uniform probability of failure for members with different numbers of bars, the values of γ_m for ten and fifty bars crossing the failing section would be 0·97 and 0·91 respectively.

To determine these values Beeby assumed a normal distribution for the strengths of individual bars and, based on test data, a standard deviation for the yield strengths of individual bars of 10% of the characteristic strength. On the basis of these assumptions, it can be seen that the capacity of a slab could be assessed on the assumption of a design strength up to 20% higher than for a simply supported beam with only two bars, and still have the same probability of the steel yielding under the design load.

3.5.5 Shear Strength

The mechanics of shear failure in reinforced concrete slab bridges are not well understood. In view of the complexities and the many parameters involved, codes of practice express shear capacities in terms of nominal average shear stresses acting over nominal cross-sectional areas. Once an area has been specified, the appropriate shear stress can be determined from test data. A fuller discussion of the factors affecting shear capacities and a comparison of design methods has been presented by Cope and Clark.[15]

Shear in reinforced concrete slab bridges is currently considered at the ultimate limit state. Two categories of shear failure are covered by codes of practice. These are beam, or flexural–shear, and punching. The terms and categories are taken from codes of practice that were drafted with building slabs, rather than slab bridges, in mind. Because building slabs tend to be uniformly loaded and to have regular plan forms, shear failure of whole slabs acting essentially as a number of parallel beam strips is readily envisaged. In such circumstances, it is reasonable that code clauses based on tests of beams should also be used for slabs. However, slab bridges are often built with skewed or curved plan forms, and the distribution of shear forces can vary quite rapidly on trial cross-sections. In addition, in such slabs, the directions of maximum shear force, principal moments and reinforcing bars can all be different. As the distribution of applied shear forces and the major parameters governing the mechanism of internal resistance are both different to those in a beam, design clauses based on beam tests should be used with caution when assessing a slab bridge. Codes leave the section width over which shear forces are averaged to the judgement of the engineer.

Punching design clauses in building codes are based on consideration of slabs supported by, and integral with, columns and of isolated, concentrated loading acting in the interior of a slab. The methods of design implicitly assume widely spaced columns or loads. Edge bearings and intermediate supports of slab bridges are usually relatively close together, with respect to the slab depth, so that design punching surfaces can intersect. This is particularly the case in British practice, which uses an unusually large offset from the loaded area to the perimeter of the failure surface considered in design.

The development of code recommendations for shear strength evaluation in North America has been reviewed by Klein and Popovic.[24] They report that, for typical concrete strengths and reinforcement ratios, the shear stresses permitted by the current code[3] are up to 40% less than those allowed in 1949.

Prior to about 1973, most slab bridge designs to British codes used a permissible working shear stress of about $0.9 \, \text{N/mm}^2$. British Standard CP 114[30] required no shear reinforcement when the calculated working shear stress was less than $U_w/50 + 0.27$ and $0.9 \, \text{N/mm}^2$. The depth of the design shear section was taken as being equal to the lever arm. When the permissible shear stress was exceeded, shear reinforcement was required to resist the entire shear force.

Design for shear was considerably revised with the introduction of limit state methods. The main effect of the changes was to reduce the permissible working shear stress on the old design cross-section to about 0.5–$0.6 \, \text{N/mm}^2$. However, when shear reinforcement was provided, a part of the shear force could be considered to be resisted by the concrete. Flexural shear and punching continued to be assessed separately for both building and bridge slabs, and no guidance was given in the Department of Transport Technical Memorandum (Bridges) BE 1/73 (1st Revision, 1979)[31] for those situations in bridge decks that did not readily fit into those categories. The above document adopted the reduced shear capacities of the limit state code CP 110, but required calculations to be carried out under working load conditions and retained the lever arm as the depth of the design section.

Ultimate limit state design for slab bridges against shear failure was introduced to the UK by BS 5400.[32] The design rules were based on those in CP 110, which were in turn based on the work of the Shear Study Group of the Institution of Structural Engineers.[33] The latter studied flexural shear failure of beams. The data from beams without shear reinforcement indicated that the main parameters governing shear failure were the percentage of tensile reinforcement, the concrete cube strength and the shear span a_v. Because of the relatively high permissible shear stresses in CP 114 and because BS 5400 does not require minimum stirrups to be provided for slab bridges unless more than 1% compression steel is present, there are many slab bridges in existence with little, if any, shear reinforcement.

The tensile reinforcement of a member contributes to the shear capacity of a section directly by dowel action, and also by controlling the crack widths. By limiting the crack widths, the steel enhances the shear capacity by enabling shear to be transferred by interlock of aggregate particles across cracks. The steel area to be considered in an assessment of shear capacity is, therefore, that located at the intersection of the most probable shear crack with the bars. To be effective, the steel should be adequately anchored beyond that critical intersection. Any deterioration of the concrete or reinforcement which is likely to affect the dowel or bond actions (for example cracking parallel to the bars, such as may have occurred with alkali–aggregate reaction or chloride-induced corrosion and spalling) should be taken into account when assessing the effective area of tensile reinforcement. The inclination of the bars to the plan direction of the shear crack should also be taken into account when assessing the effective steel area, as the effectiveness of the reinforcing bars depends on their stiffnesses to opening of the shear crack.[15]

The concrete compressive strength influences the portion of the shear capacity due to the uncracked concrete 'above' the shear crack. Any deterioration in this zone should be taken into account when assessing the shear capacity.

Taylor[34] has suggested that the contributions to the section capacity of a beam due to aggregate interlock, dowel action and shear in the compression zone could lie in the ranges 33–50%, 15–25% and 20–40% respectively. These values were obtained by combining the results of studies of dowel capacity, aggregate interlock action and beam failures. The values should be treated with caution, however, as it is difficult to reproduce with accuracy the complex stress state of a cracked, reinforced member. Recent research into shear failure of beams without shear reinforcement suggests that the aggregate interlock contribution may not be as significant. This is because the displacements at ultimate have been found to be almost purely rotational, with little shear deformation.[35]

Test data from beam studies indicate that the unit shear strength of a member decreases with increasing depth. This effect was not taken into account in CP 110, because building slabs are relatively thin and because the deeper members (beams) are provided with minimum stirrups. Taylor[36] considered that the provision of minimum stirrups in beams would effectively take care of any reduction in the concrete shear capacity with depth. BS 5400: Part 4 extends the range of the depth factor, ξ_s, to reduce the concrete contribution to the shear capacity in slabs more than 500 mm deep. The effect of the depth factor is not taken into account in the United States.[3]

It is a feature of tests on beams and slab–column junctions that the nominal ultimate shear stresses show great scatter. Because test data are limited, the actual material partial safety factors adopted for shear are not always obvious. This is unfortunate as, due to considerable reductions of code shear strengths over the years, many existing slab bridges may not satisfy current code requirements.

3.5.5.1 Section Capacity

BS 5400 gives the concrete contribution to the shear capacity of a section as

$$v_c = \frac{0.27}{\gamma_m} \left(\frac{100 A_s}{b_\omega d}\right)^{1/3} (f_{cu})^{1/3} \text{ N/mm}^2 \tag{3.6}$$

The first expression in parentheses is the effective percentage of well-anchored tensile, flexural reinforcement normal to the shear failure plane, but may not take a value greater than 3. The term in the second parentheses is the cube strength (characteristic), but a maximum value of 40 N/mm^2 may be used in the formula. γ_m is taken as 1·25.

The concrete shear strength given by eqn (3.6) is modified by a depth factor, which is less than one for effective depths greater than 500 mm. Equation (3.6) is the basis of the tabulated shear strengths in CP 110, BS 8110 and BS 5400. However, there remains some controversy over whether setting γ_m to 1·25 gives the characteristic or design strength of members with no shear reinforcement.

Allen[37] has studied the data and recommendations of the Shear Study Group of the Institution of Structural Engineers[33] and gives the ultimate shear capacity equation for beams with shear reinforcement as

$$v_{ult} = k v_c + 1\cdot 1 \frac{A_{sv}}{b s_v} f_{yv} \tag{3.7}$$

Shear strengths given by this equation are shown in Fig. 3.6 as a continuous line. It can be seen that the values given by the formula could be used as characteristic strengths until more data become available.

The design shear strength recommended by the Shear Study Group was obtained by dividing v_{ult} by 1·25 ($\gamma_m = 1\cdot 25$). However, Allen says that 'After further consideration, it was felt that the 0·8 factor (1/1·25) was too restrictive, so the concrete contribution was increased to its full value'. The design shear strengths given by the revised design equation are shown in Fig. 3.6 as the line of short dashes. Effectively, for a member with no shear reinforcement, substitution of 1·25 for γ_m in eqn (3.6) gives a material partial safety factor of only 1. The actual partial safety factor for a member with shear reinforcement increases with the area of shear reinforcement.

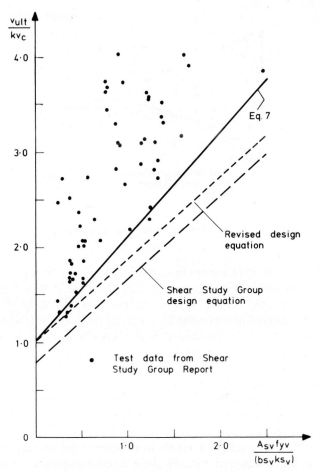

Fig. 3.6.   Shear equations (based on ref. 37).

Figure 3.7 compares the design shear strengths of BS 5400 (1984) with test data for beams without shear reinforcement gathered by the Shear Study Group. The figure is based on Fig. 2 of Baker, Yu and Regan.[38] In their paper, the authors state that the test data are for beams with shear span/effective depth ratios >3, concrete with cube strengths in the range 30–37·5 N/mm², and in which shear failures were sudden and occurred at load levels close to those initiating shear cracking. Failures, therefore, occurred with little warning.

R. J. COPE

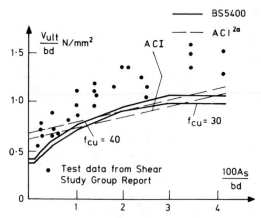

Fig. 3.7.   *Design shear strengths compared with test data.*

For beams, where minimum links have traditionally been provided, the partial safety factor is always greater than 1, but for slabs without shear reinforcement, that behave as wide beams, it would seem prudent to introduce a material partial safety factor on the concrete shear strength. An appropriate safety factor for shear could lie between the values for steel and concrete, as the shear capacity is dependent on the tension steel area and the concrete cube strength. Setting $\gamma_m = 1\cdot 5$ in eqn (3.6) would provide a material partial safety factor of $1\cdot 2$.

The test data used to obtain the v_c values for use in a design or assessment of a slab bridge were based on tests of beams. The stress conditions in skewed and curved slabs are considerably more complex than those in beams, as flexural cracks can initiate from both the top and soffit surfaces and both horizontal and vertical shear stress components have to be transmitted. Test data for shear failure in irregular slabs are few, and it is not possible to state the actual partial factor of safety. It is suggested, therefore, that increasing the value of γ_m in eqn (3.6) to $1\cdot 5$ be considered as a minimum increase, and that higher values be considered when high shear force intensities are accompanied by large twisting moments.

Figure 3.7 includes the design values for the current American code.[3] The lack of conservatism for tensile steel percentages of less than 1% is well known.[39] Because minimum shear reinforcement is provided in beams, safe designs result. However, for slab bridges without shear reinforcement, the tensile steel in the zones critical for shear may be quite low due to bar curtailment. Use of the modified shear design equations recommended in ref. 36 should, therefore, be considered in preference to those given in the code for zones of slabs with steel percentages of less than 1%.

The test data used to obtain the v_c values were obtained from beams with non-curtailed reinforcement placed in the bottom of the beam. Clark and Thorogood[40] have conducted tests and reviewed test data of others on the shear strength of beams with tension bars cast in the top of the member. They reported reductions of shear strength and greater scatter of experimental data. The main reason for these differences is believed to be the poorer bond in the less well compacted concrete. However, CIRIA[1] reports that the mean position of top-placed bars in slabs cast on sites is systematically lower than the specified position, so the effective depth is likely to be less than that assumed when calculating the shear capacity of a section. Also, if top bars are simultaneously stressed to yield by flexural action, they would have little stiffness remaining to control the width of shear cracking. Baron[41] has shown that the shear strength of a member can be reduced by the presence of curtailed reinforcement. The stress concentrations at the cut-off point cause cracking at relatively low loads, which leads to a redistribution of stresses and thus to a lower failure load. When considering the possible effect of curtailed reinforcement, the intersection of the reinforcement with the likely shear crack and not with the design shear plane should be considered.

There are insufficient data to enable a confident determination of the characteristic shear strength for these situations. Until there is, it would seem prudent to increase the partial safety factor by at least 30% when the tension steel is in the top of a slab and when a critical section is in a region of bar curtailment.

In European practice, the basic allowable shear stress, v_c, is factored by ξ_s, which is a function of the effective depth:

$$\xi_s = (400/d)^{1/4} \geq 1 \cdot 0 \qquad \text{BS 8110 (1985)} \qquad (3.8a)$$

$$\xi_s = (500/d)^{1/4} \geq 0 \cdot 7 \leq 1 \cdot 50 \qquad \text{BS 5400 (1984)} \qquad (3.8b)$$

The major difference between these recommendations is that the bridge code factor reduces the shear capacity for slab bridges with effective depths greater than 500 mm. This is consistent with experimental trends (see Fig. 3.8). The figure uses data collated by Clark.[8] The reason why the building code does not reduce the shear capacity with increasing depth may be that building slabs are rarely as deep as 400 mm and shear reinforcement would be provided in beams. However, there are few data for deep slabs, and the reduction may be pessimistic. Test data for thick slabs subjected to the

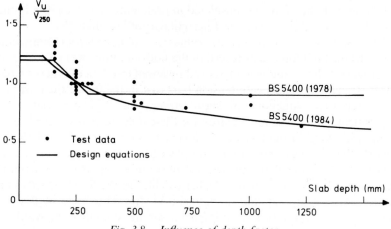

Fig. 3.8.   Influence of depth factor.

shear stress conditions encountered in slab bridges are urgently required to provide realistic values for the material partial safety factor.

The third main parameter governing shear capacity identified by the Shear Study Group[33] was the shear span. For short shear spans, the basic allowable shear stress may be enhanced by a factor proportional to the ratio of the effective depth to the shear span. The factor recommended by BS 5400 for beams is $2d/a_v$, where a_v is the shear span. However, the short shear span enhancement factor is not referred to in the code clauses dealing with the shear resistance of slabs. In American practice,[3] the maximum design shear force considered is that acting on the sections at an effective depth from the face of a support. BS 8110 recommends that the enhancement be used, for beams, in any situation where the section considered is closer than twice the effective depth to the face of a support or a concentrated load. In a separate clause, a procedure is suggested in which the enhancement is taken into account indirectly, by considering shear forces only at sections located at distances more than an effective depth from the face of a support. Only this clause is referred to when the flexural–shear resistance recommendations are given for solid slabs. However, for punching shear, v_c may be increased by the factor $1.5d/a_v$ when perimeters closer than $1.5d$ from the loaded area are considered.

A comparison of the available test data with the recommended enhancement formula for rectangular beams without shear reinforcement has been provided by the Shear Study Group.[33] It can be seen from Fig. 3.9 that the formula gives conservative predictions and that the actual material

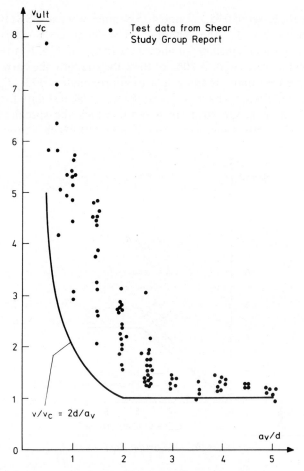

Fig. 3.9. Influence of shear span (based on ref. 37).

partial safety factor is greater than that stated in the code. However, most of the tests used concentrated loadings and there was no transverse distribution of shear force as occurs in slab bridges. Until more test data become available, it is not possible to state the actual material safety factor appropriate to slab bridges supported on discrete columns or bearings.

If the shear resistance of the concrete is exceeded, shear reinforcement may be provided, subject to a specified maximum shear stress not being exceeded. This condition is imposed to prevent crushing of the concrete parallel to the shear cracks. BS 5400 specifies that the maximum shear stress should be the lesser of $0.75\sqrt{f_{cu}}$ and $4.75\,\text{N/mm}^2$, whereas BS 8110 specifies the lesser of $0.8\sqrt{f_{cu}}$ and $5\,\text{N/mm}^2$. The latter code states that its recommended limits include an allowance for γ_m of 1·25. If it is assumed that the cylinder strength is 80% of the cube strength, the corresponding North American limit[3] is equivalent to approximately $0.73\sqrt{f_{cu}}$.

Clarke and Taylor[42] have reviewed the available test data for crushing failure of the concrete struts in a beam, which the specification of a maximum shear stress aims to prevent. Figure 3.10, which is based on Fig. 5

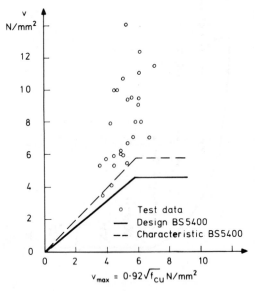

Fig. 3.10.   *Maximum shear stress.*

of their report, compares the recommendations of BS 5400 for reinforced concrete with the test data. The characteristic line for BS 5400 has been obtained by multiplying the design values by the partial safety factor $\sqrt{\gamma_m} = \sqrt{1.5} \approx 1.22$. In view of the large scatter in the test data and the limited number of data points, it is reasonable to use the recommendations of BS 5400 to obtain a partial safety factor of about 1·2, which is consistent with setting γ_m to 1·5 in eqn (3.6) for the shear resistance of the concrete.

Figure 3.6 compares test data assembled by the Shear Study Group with design equations for members with shear reinforcement prior to the introduction of BS 5400. The Shear Study Group did not examine the effects of repeated loading on shear capacity. In BS 5400: Part 4,[2] the calculated concrete contribution, v_c, to the shear capacity of a beam or slab with shear reinforcement is required to be reduced by $0.4\,\text{N/mm}^2$. For a typical slab bridge, this is equivalent to reducing v_c by 50–60%. Clarke and Pomeroy[43] have suggested that the reduction is intended to allow for the reduced effect of aggregate interlock under repeated loading. However, if this is the case, it has presumably been assumed that shear cracking is present under working conditions. If there is no shear cracking under working loads, there can be no loss of the aggregate interlock contribution to the shear capacity due to fatigue. As no test data are known to justify the reduction, it is not possible to quantify its effect on the partial safety factor of a section with shear reinforcement. For the assessment of slab bridges the difficulty is largely academic as few older structures contain shear reinforcement. The reduction is not considered for assessments of the punching shear resistance of a slab.

Klein and Popovic[24] have reviewed the literature on the fatigue strength of beams without shear reinforcement, and they report tests[44] which show that the fatigue strength can be substantially less than the static strength. Their paper gives fatigue reduction factors (FRF) with the number of cycles to failure. However, in the tests studied, the shear stress range was almost 100% of the maximum shear stress. In slab bridges, the dead load can be substantial and a formula to give an adjusted fatigue reduction factor (FRF'), based on the test results of Aas-Jacobsen[45] using different ratios of dead load shear force to total load shear force (β_d), is also given. For 1000, 10 000 and 100 000 load cycles, the fatigue reduction factors are approximately 0.87, 0.7 and 0.66 respectively. The adjusted fatigue reduction factor is given by $\text{FRF}' = 1 - (1 - \text{FRF})(1 - \beta_d)$.

To enable shear reinforcement to function, longitudinal tensile reinforcement is required. This reinforcement is in addition to that required for bending. In the Building Code CP 110, a minimum amount of longitudinal reinforcement (of the order 30–50%) is carried through to the supports, and this probably ensures that there is always sufficient reinforcement to resist both bending and shear. In BS 5400: Part 4, the minimum area of effectively anchored longitudinal reinforcement required in the tensile zone is given by $A_s = V/(2f_y/\gamma_m)$. V is the shear force due to ultimate loads at the section considered. However, Clark[46] has emphasised that this should be the minimum amount of tensile reinforcement required to act with vertical shear reinforcement, and is in addition to that required to resist bending. For an appraisal, it is suggested that a check be made to ensure that there is sufficient additional, well-anchored tensile reinforcement provided to enable the shear reinforcement to function as intended.

3.5.5.2 Punching

For a concentrated load or reaction away from an edge, BS 5400: Part 4[2] specifies the perimeter of the design shear section to be rectangular and to be at 1·5 times the effective depth from the loaded area. This perimeter is similar to that specified in CP 110, but is much further from the loaded area than the perimeters considered in CP 114 and the American and European codes. The actual punching shear failure surface is not considered, and the design surface was selected to enable the same average shear stress formula, eqn (3.6), to be used for the assessment of both flexural and punching shear capacities.

Regan[47] has reviewed the test data for punching shear at internal slab–column connections under concentric loading, and gives the characteristic strength as

$$V_k = 0·27k\left(\frac{300}{d}\right)^{1/4}\left(\frac{100A_s}{bd}\right)^{1/3}(f_{cu})^{1/3}d(1·3c + 10·2d) \qquad (3.9)$$

where k is a factor for the influence of the shape of the contact area and is 1·02 for a square, 1·15 for a circle and 0·96 for a rectangle with an aspect ratio of 2; c is the perimeter of the loaded area in millimetres.

The design punching shear strength according to BS 5400 is

$$V_d = \frac{0·27}{\gamma_m}\left(\frac{500}{d}\right)^{1/4}\left(\frac{100A_s}{bd}\right)^{1/3}(f_{cu})^{1/3}d(c + 12d) \qquad (3.10)$$

where c is the perimeter of a rectangular column or the perimeter of the square enclosing a circular column, and $\gamma_m = 1·25$.

For a typical slab bridge with an effective depth of 700 mm, supported on circular contact areas with diameters of 900, 700 and 200 mm, the partial factors of safety obtained by dividing eqn (3.9) by eqn (3.10) are 1·14, 1·13 and 1·09 respectively. Thus, as for flexural shear with no shear reinforcement, the actual partial factor of safety appears to be less than the 1·25 specified in the code and, for an appraisal, it would be prudent to set γ_m to 1·5 rather than to 1·25 as specified. This would result in a material partial factor of safety of about 1·3 for relatively small contact areas, which is consistent with the partial safety factor for punching shear recommended by Regan.[47]

In Regan's formula the average steel percentage for the two orthogonal steel directions is taken, whereas in the BS 5400 formula the steel percentages appropriate to each part of the failure perimeter are used to determine the design shear stresses on the associated cross-sectional areas.

The effect of this is to increase the partial factor of safety by a few per cent for highly orthotropic steel arrangements.

Regan's[47] comprehensive survey of international test data encompasses rectangular supports with aspect ratios up to 4; different concentrations of flexural reinforcement in the critical zone; different strengths of the flexural reinforcement; the effect of compression reinforcement; and the effect of predominantly one-way bending. However, it should be emphasised that Regan reported only one test result for an effective slab depth greater than 200 mm and no eccentricity of loading was considered. It has been assumed that the effect of depth on punching shear strength is the same as that on flexural shear strength.

When the concrete shear resistance is exceeded, part of the shear force can be resisted by shear reinforcement, provided a minimum quantity of shear reinforcement is present. The presence of this minimum quantity is required to ensure that the cracked section is at least as strong as the uncracked section. BS 5400 specifies the minimum area of shear reinforcement to be

$$A_{sv} = 0.4 \sum (bd)/(f_{yv}/\gamma_{ms}) \tag{3.11}$$

For design purposes, the code requires the shear reinforcement to be determined for a perimeter $1.5d$ from a column face and then at successive perimeters progressively $0.75d$ out from the critical perimeter. However, the calculated reinforcement on the innermost design perimeter must also be provided on a parallel perimeter $0.75d$ inside it. This is to prevent a steep shear crack 'avoiding' the shear reinforcement. BS 8110 also requires reinforcement close to the column, but only half of the quantity determined by the above equation need be provided on each design perimeter. This is because tests have shown that the failure cracks intersect the shear reinforcement on two perimeters.

Regan[47] has reviewed the test data for slabs with shear reinforcement, and has concluded that there is a lack of reliable results. It is not surprising, therefore, that code recommendations differ and that the actual partial safety factor cannot be determined with any degree of precision.

CP 110 originally required the shear reinforcement for an internal column to be determined from

$$A_{sv} = (V_d - 0.8\xi_s v_c bd)/(f_{yv}/\gamma_{ms}) \tag{3.12a}$$

where V_d is the shear force transferred from the slab at the ultimate limit state. BS 5400[2] gives the same recommendation for columns near an edge of a slab, but does not include the 0.8 factor for interior columns. This is believed to be because the code drafters felt that with no moment transfer, the effects of a non-uniform shear distribution need not be considered. Regan[48] has observed, however, that the test results for slabs with an integral column or a concentrated load are very similar.

In a revision to CP 110, eqn (3.12a) was revised to

$$A_{sv} = (1 \cdot 25 V_d - \xi_s v_c bd)/(f_{yv}/\gamma_{ms})$$ (3.12b)

In the discussion of γ_{f_3} values earlier, it was argued that this formula is more logical, and that the factor $1 \cdot 25$ should be considered as a γ_{f_3} partial safety factor, to account for the uncertain, uneven shear distribution around a design perimeter. For an assessment, it is therefore suggested that the punching resistance on a design perimeter of a column with no moment transfer could be based on

$$V = \left(\frac{\xi_s v_c bd}{\gamma_{mc}} + \frac{A_{sv} f_{yv}}{\gamma_{ms} = 1 \cdot 15}\right)$$ (3.13)

where $\gamma_{mc} = 1 \cdot 2$ when there is no shear reinforcement and unity when at least the code-specified minimum quantity of such reinforcement has been provided. The effective shear force used with this equation would then be the column reaction predicted by a linear elastic analysis factored by a γ_{f_3} value of $1 \cdot 25$, to take into account the effects of a non-uniform shear distribution. In view of the recommendations of BS 8110, it is suggested that the area of shear reinforcement considered be that in a failure zone, and that would normally encompass the shear steel on the perimeter under consideration and on the neighbouring perimeter closer to the column.

When moment is transferred to a column, it is suggested that the calculated applied shear force be replaced by an effective shear force according to the recommendations of BS 8110, provided that the effective shear force is at least 25% greater than the calculated shear force.

An estimate of the material partial safety factor, when there is moment transfer at an internal column or at an edge column with moments parallel to the slab edge, can be obtained by comparing the calculated shear resistance with the value from the formula given by Regan:[47]

$$V_{ult} = V_k/\{1 + 2e/\sqrt{[(c_1 + 2d)(c_2 + 2d)]}\}$$ (3.14)

In eqn (3.14), V_{ult} is the shear capacity with moment transfer; V_k is the ultimate concentric shear capacity given by eqn (3.9); e is the ratio of the moment and the force transferred to the column; c_1 and c_2 are the column dimensions. Regan[48] has shown that the values given by eqn (3.14) give a good lower bound to the available test data. However, it must be stressed

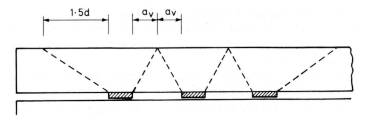

Fig. 3.11. Overlap of punching perimeters.

that the test data are very limited, and the reader is advised to consult Regan's report for details.

Where a slab bridge is supported on a series of discrete bearings, critical perimeters for punching specified by British codes may overlap (see Fig. 3.11). In such situations, it is suggested that punching of groups of bearings together be considered, but punching by individual bearings should also be taken into account. This can be particularly important for highly skewed slabs, when a large portion of the reaction may fall on the outermost bearing in the obtuse corner.[16] When a single bearing is considered, the smaller perimeters and the higher design shear stresses of the CEB Model Code or the ACI Code[3] could be considered. Alternatively, the design perimeter specified in BS 5400 could be modified, as indicated in Fig. 3.11, and the shear stress on the portion of the perimeter closer than $1 \cdot 5d$ to the loaded area increased by a factor $1 \cdot 5d/a_v$.

Tests on scale models of skew slab bridges supported on lines of discrete bearings[16] have shown the possibility of shear failure involving only the most heavily loaded group of bearings. The paper cited provides some guidance on critical sections to be considered for an assessment of shear capacities of slabs with skews of 30°, 45° and 60°.

3.6 COMMENTS

Existing reinforced concrete slab bridges showing no signs of distress may not satisfy current design codes. Various explanations of their survival are possible. A deck may not have been subjected to the design loading. (The loading may be unrealistic for the bridge location, or have too low a probability of occurrence for a bridge in any location.) Alternatively, critical section capacities may be greater than the calculated values, which may not have been based on moments and shear forces averaged over sufficiently long sections. By taking into account the bases of current code stipulations and test data, an engineer would be in a better position to make rational assessments of each particular problem than if code procedures were followed blindly.

One of the major difficulties is to determine the factor of safety against shear failure in slabs with no shear reinforcement. When punching is considered, a slab is assumed to possess sufficient ductility for the shear stresses to be averaged over the design shear perimeter. In British practice, this is at $1·5d$ from the face of a support. This perimeter has a considerable length in comparison to the slab width and could have large variations of shear forces predicted by a linear analysis. However, when 'beam or flexural' shear is considered, the section length over which the predicted shear force is to be averaged is not specified. For skewed slabs, the shear force intensity can vary rapidly in the obtuse corner region, and only limited test data are available[16] to give guidance on the selection of the appropriate section length over which shear stresses may be averaged. More test information is urgently required.

For many bridges, the regulating authority specifies the loading, suggests acceptable methods of analysis and specifies procedures for determining resistances. However, if prescribed procedures cannot reasonably be used, due to poor detailing or deterioration of the structure, an engineer is free to use engineering judgement. In this chapter, a critical review of current methods and of the data on which they are based has been provided to assist with such cases.

When a structure has deteriorated, assessment of strength is particularly difficult. To provide rational estimates, the effects of the deterioration on the various factors contributing the strengths need to be determined. For this purpose, the European design philosophy, with partial factors of safety applied to the different materials, may be more useful than the American practice of applying single reduction factors to section strengths.

If corrosion has affected the ductility of the reinforcement, only limited moment redistribution can be considered and the use of the yield line method would need careful assessment. If corrosion has reduced the cross-section of the reinforcement, the material partial safety factor for the steel could be increased to account for the additional uncertainty introduced. If corrosion-induced spalling has occurred, the shear capacity is likely to be reduced not only by the loss of concrete cross-section but also by reductions in the contribution of the dowel action, and in the aggregate interlock capacity due to the reduced ability of the reinforcement to control crack widths. Aggregate interlock capacity could also be reduced by water penetration causing leaching of chemicals in the concrete paste at the crack interfaces.

The effects of alkali–silica reaction on section capacities have yet to be quantified. To provide rational assessment procedures for assessing bending and shear capacities, its effects on bond, dowel action, aggregate interlock capacity and the shear capacity of the compression block are urgently required. When such data are available, they could be used with the information presented in this chapter to provide a rational means of assessing the strength of affected slab bridges.

REFERENCES

1. Construction Industry Research and Information Association, *Rationalisation of safety and serviceability factors in structural codes*, CIRIA Report 63, London, p. 226 (1977).
2. British Standards Institution, BS 5400, *Steel, concrete and composite bridges: Part 4. Code of Practice for concrete bridges*, London (1984).
3. American Association of State Highway and Transportation Officials, *Standard Specification for highway bridges*, 13th edn (1983).
4. Institution of Structural Engineers, *Appraisal of existing structures* (1980).
5. Department of Transport, BD 21/84, *The assessment of highway bridges and structures* (1984).
6. British Standards Institution, BS 5400, *Steel, concrete and composite bridges: Part 2. Specification for loads*, London (1978).
7. Henderson, W., British highway bridge loading, *Proc. ICE*, Part 2, **3**, 325–73 (1954).
8. Clark, L. A., *Concrete Bridge Design to BS 5400*, Construction Press, London (1983).
9. Flint, A. R. and Edwards, L. S., Limit state design of highway bridges, *Struct. Engineer*, **48**(3), 93–108, and **48**(9), 372 (1970).
10. Flint, A. R., Smith, B. W., Baker, H. J. and Manners, W., The derivation of safety factors for design of highway bridges, in: *The Design of Steel Bridges*, ed. Rockey, K. C. and Evans, H. R., Granada, London, 11–36 (1981).
11. Armitage, A., Report of the inquiry into lorries, people and the environment, HMSO, London (1980).
12. Statutory Instruments: *Motor vehicles construction and use regulations* (SI 1978/1017) (1978) and *Motor vehicles (construction and use) (Amendment) (No. 7) regulations* (SI 1982/1576) (1982) and *The motor vehicles (authorisation of special types) general order* (SI 1981/859) (1981).
13. Department of Transport, BD 14/82, *Loads for highway bridges. Use of BS 5400: Part 2: 1978* (1982) and *Amendment No. 1*, p. 2 (1983).
14. Church, J. G. and Clark, L. A., Combination of highway loads and temperature difference loading on bridges, *Struct. Engineer*, **62A**(6), 177–81 (1984).
15. Cope, R. J. and Clark, L. A., *Concrete Slabs: Analysis and Design*, Elsevier Applied Science, London (1985).

16. Cope, R. J., Flexural shear failure of reinforced concrete slab bridges, *Proc. ICE*, Part 2, **79**, 559–83 (1985).
17. Cope, R. J. and Rao, P. V., Shear forces in edge zones of concrete slabs, *Struct. Engineer*, **62A**(3), 87–92 (1984).
18. Cope, R. J. and Rao, P. V., Non-linear finite element analysis of concrete slab structures, *Proc. ICE*, **53**(2), 159–79 (1977).
19. Cope, R. J. and Rao, P. V., Moment redistribution in skewed slab bridges, *Proc. ICE*, **75**(2), 419–51 (1983).
20. Cope, R. J. and Rao, P. V., A two-stage procedure for the non-linear analysis of slab bridges, *Proc. ICE*, **75**(2), 671–88 (1983).
21. British Standards Institution, CP 110, *The structural use of concrete: Part 1. Design, materials and workmanship*, as amended by AMD 1553 (1974) and AMD 1881 (1976), London (1972).
22. British Standards Institution, BS 8110: Part 1, *Structural use of concrete: Part 1. Code of Practice for design and construction*, London (1985).
23. Cope, R. J., Non-linear analysis of reinforced concrete slabs, Chapter 1 in: *Computational modelling of concrete structures*, ed. Hinton, E. and Owen, D. R. J., Pineridge Press, pp. 1–41, Swansea (1986).
24. Klein, G. J. and Popovic, P. L., Shear strength evaluation of existing concrete bridges, in: *Strength Evaluation of Existing Concrete Bridges*, ACI, SP-88, 199–224 (1985).
25. British Standards Institution, BS 8110: Part 2, *Structural use of concrete: Part 2. Code of Practice for special circumstances*, London (1985).
26. McGowan, R. and McCafferty, J. P., P1/P2 interchange, Hong Kong, *Struct. Engineer*, **63A**(10), 318–26 (1985).
27. Bungey, J. H., *The Testing of Concrete in Structures*, Surrey University Press (1982).
28. Sturman, G. H., Shah, S. P. and Winter, G., Effect of flexural strain gradient on micro-cracking and stress–strain behaviour of concrete, *ACI J.*, **62**(7), 805–22 (1965).
29. Beeby, A. W., *A proposal for changes to the basis for the design of slabs*, Tech. Rep. 547, Cement and Concrete Association, Slough (1982).
30. British Standards Institution, CP 114, *The structural use of reinforced concrete in buildings* (1969).
31. Department of Transport, BE 1/73, *Reinforced concrete for highway structures* (1979).
32. British Standards Institution, BS 5400, *Steel, concrete and composite bridges: Part 4. Code of Practice for concrete bridges*, London (1978).
33. Shear Study Group of the Institution of Structural Engineers, *The shear strength of reinforced concrete beams*, IStructE, London (1969).
34. Taylor, H. P. J., The fundamental behaviour of reinforced concrete beams in bending and shear, in: *Shear in Reinforced Concrete*, Vol. 1, ACI SP-42, American Concrete Institute, Detroit, 43–77 (1974).
35. Chana, P. S., Private communication, Cement and Concrete Association (1986).
36. Taylor, H. P. J., Shear strength of large beams, *Proc. ASCE*, **98**(ST11), 2473–90 (1972).
37. Allen, A. H., *Reinforced concrete design to CP 110*, Cement and Concrete Association, London (1977).

38. Baker, A. L. L., Yu, C. W. and Regan, P. E., Explanatory note on the proposed Unified Code clause on shear in reinforced concrete beams with special reference to the report of the Shear Study Group, *Struct. Engineer*, **47**(7), 285–93 (1969).
39. ACI–ASCE Committee 426, Suggested revision to shear provisions for Building Codes, *ACI J.*, **74**(9), 458–69 (1977).
40. Clark, L. A. and Thorogood, P., Shear strength of concrete beams in hogging regions, *Proc. ICE*, Part 2, **79**, 315–26 (1985).
41. Baron, M. J., Shear strength of reinforced concrete beams at points of bar cut-off, *Proc. ACI*, **63**(1), 127–34 (1966).
42. Clarke, J. L. and Taylor, H. P. J., *Web crushing—a review of research*, Tech. Rep. 42,509, Cement and Concrete Association, London (1975).
43. Clarke, J. L. and Pomeroy, C. D., Concrete opportunities for the structural engineer, *Struct. Engineer*, **63A**(2), 45–53 (1985).
44. Chang, T. S. and Kesler, C. E., Static and fatigue strength in shear of beams with tensile reinforcement, *Proc. ACI*, **54**, 1033–57 (1957).
45. Aas-Jacobsen, K., *Fatigue of concrete beams and columns*, Norwegian Institute of Technology, Trondheim, Bulletin No. 70–1.
46. Clark, L. A., Longitudinal shear reinforcement in beams, *Concrete*, **18**(2), 22–3 (1984).
47. Regan, P. E., *Behaviour of reinforced concrete flat slabs*, CIRIA Report 89 (1981).
48. Regan, P. E., Design for punching shear, *Struct. Engineer*, **52**(6), 197–208 (1974).

4

Repair and Protection of Reinforced Concrete Bridges

PAUL BENNISON

Liquid Plastics Ltd, Preston, UK

4.1 INTRODUCTION

Since the first reinforced concrete bridge was constructed in the UK in 1902, a further 50 000 bridges have now been constructed in either reinforced or prestressed concrete. Like all concrete structures, concrete bridges give excellent durability when designed, constructed and maintained correctly, justifying the design life of 120 years for such a structure. However, as the basic design principles of both reinforced/prestressed concrete and construction were refined, more elegant and complex bridge structures were built with subsequent enhanced aesthetics and economy. Unfortunately, this also led to serviceability problems with some bridges and brought into question the durability of concrete, which is relatively cheap and still the most widely used construction material in the world.

The realisation that modern concrete bridges are not as durable as was anticipated has caused increasing concern amongst structural design engineers, site contractors and material scientists alike. It is now accepted that repair procedures for defective concrete bridges should be approached with the same technical expertise as was the original design in order to achieve the most effective engineering solutions.

4.2 DURABILITY OF CONCRETE

The quality of the base concrete used in concrete bridges is of great importance; concrete by its very nature is a long-lived material, many examples dating back to Roman times and still in good condition. Unfortunately, several factors govern its life in service and, therefore, bridge concrete must be of the highest quality, i.e. as durable and impermeable as possible. Durable, low permeability concrete is synonymous with good quality, dense, well-compacted concrete.

The following factors will significantly affect the durability of concrete:

(1)   the mix design;
(2)   the cement content and type;
(3)   the water/cement ratio;
(4)   the degree and type of compaction;
(5)   the quality and type of the aggregates;
(6)   curing;
(7)   the use of chemical admixtures.

The design engineer should also consider how he can influence the permeability of the concrete in bridges by making allowances for primary and secondary stresses liable to cause cracking.

4.3   POROSITY AND PERMEABILITY

For a cement matrix having a coherent pore system (porosity) to become permeable, there must be some form of interconnecting system of tubes or canals. Obviously, discrete sealed pores will not lead to either air or water permeability (see Fig. 4.1). Also, the size and width of pores have an influence on permeability, because the narrower the pores, the higher must

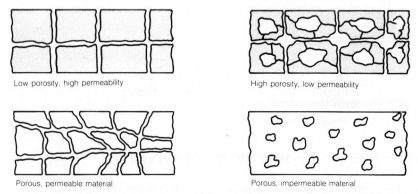

Low porosity, high permeability

High porosity, low permeability

Porous, permeable material

Porous, impermeable material

Fig. 4.1.   Illustration of permeability and porosity.[1]

TABLE 4.1
Characteristics of Gel, Capillary and Air Type Pores

Pore type	Diameter	Description	Property affected
Gel pores	10–2·5 nm	Small	Shrinkage
	2·5–0·5 nm	Micro pores	Shrinkage/creep
	<0·5 nm	Micro pores (interlayer)	Shrinkage/creep
Capillary pores	10–0·5 μm	Large	Strength/permeability
	50–10 nm	Medium	Strength/permeability, shrinkage
Air pores	0·01–0·2 mm	Very large	Strength/permeability

be the pressure required to force a liquid, generally water, through the pore system. It is important, therefore, to distinguish between air pores, capillary pores and gel pores.

Small air pores are usually artificially entrained into the matrix by chemical admixtures in order to increase workability and enhance resistance to the effects of frost. Large irregular air pores are caused by poor placing and compaction of the concrete. Capillary and gel pores result from the hydration process of cement and water (Table 4.1).

In fact, the porosity and permeability of the hardened cement paste depend upon the water/cement ratio. There is, therefore, a relationship between water/cement ratio and compressive strength, capillary porosity and compressive strength, and permeability and water/cement ratio (see Figs 4.2–4.4). It is, however, important to note that while strength and diffusion are affected by porosity, they are not uniquely interrelated.

Thus, the nature of the hardened cement can be summarised as follows. Immediately after mixing the cement and water together, an agglomeration of particles and water is formed. The water in the interconnecting cavities is termed 'capillary water'. Until it has been completely hydrated, Portland cement chemically binds water equivalent to approximately a quarter of its weight. Therefore, the water loses approximately a quarter of its volume. In addition to this chemically bound water, Portland cement loosely binds approximately 15% of its weight as 'gel water'. Despite its loose chemical bond, gel water is not able to react with unhydrated cement and, in fact, evaporates in dry air or in an oven at 105°C. It is this gel water and the additional water required to 'lubricate' the mix to form a workable material that create more capillaries as they become evaporable water. Therefore, the greater the w/c ratio, the greater the permeability (see Fig. 4.4).

PAUL BENNISON

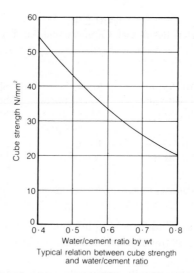

Fig. 4.2.   *Relationship between compressive strength and water/cement ratio.*[2]

The hydration by-products of the cement/water reaction form a coherent homogeneous mass, the 'cement gel'. The cement gel is, however, made up of approximately 25% by volume of finely distributed pores, namely gel pores. The total porosity (large capillary and to a lesser degree gel pores) of the hardened cement paste, therefore, determines its strength (see Fig. 4.5 and Table 4.2).

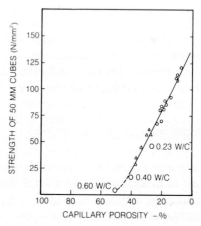

Fig. 4.3.   *Relationship between compressive strength and capillary porosity.*[3]

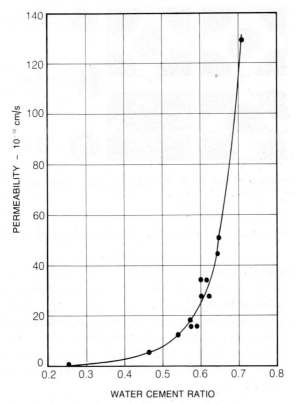

Fig. 4.4.   Relationship between permeability and water/cement ratio for mature cement pastes (93% of cement hydrated).[4]

TABLE 4.2
Constituents, Properties and Hydration Products of Cements

Cement constituents			Properties affected
C_3S	Tricalcium silicate	55%	Strength up to 28 days/setting time
C_2S	Dicalcium silicate	20%	Long-term strength, i.e. after 28 days
C_3A	Tricalcium aluminate	10%	Setting time/24 h strength
C_4AF	Tetracalcium aluminoferrite	8%	Very little effect

Main hydration products

C–S–H	Calcium silicate hydrates: form the 'gel' structure
C_3AH_6	Calcium aluminate hydrate
$Ca(OH)_2$	Calcium hydroxide: gives the cement paste its high alkalinity

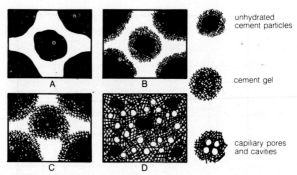

Fig. 4.5. Hydration of cement. A, Immediately after mixing; B, reaction around particles—early stiffening; C, formation of skeletal structure—first hardening; D, gel infilling—later hardening.[5]

Permeability and porosity can, therefore, be influenced in several ways, i.e.:

(1) The use of chemical admixtures:
  (a) to lower numerically the w/c ratio and reduce the amount of water in the mix;
  (b) to entrain small discrete air pores to enhance resistance to freeze/thaw cycles in vulnerable areas;
  (c) to modify the calcium silicate hydrate crystal formation during the hydration induction phase and, therefore, the gel pore formation.

(2) Curing—if concrete is poorly cured, hydration of cement is adversely affected and possible surface drying will cause plastic shrinkage cracking.

(3) The use of pozzolanic materials, which react with the calcium hydroxide liberated by the cement hydration to produce insoluble stable calcium silicate and aluminate hydrates which will block pores and capillaries. Some of the commonly used pozzolanic materials are as follows:
  (a) micro-silica;
  (b) PFA (pulverised fuel ash);
  (c) GGBFS (ground granulated blast furnace slag).

4.4 CAUSES OF DETERIORATION OF CONCRETE BRIDGES

All reinforced concrete structures crack. The skill required by the design engineer is to limit the incidence of cracks and crack widths to acceptable levels. In general terms, cracking can be defined under the two headings 'structural' and 'non-structural'. Excessive cracking (i.e. when permissible limits are exceeded) of a reinforced concrete member is caused by either a deficiency in the design or construction or, alternatively, from improper overloading. This results in either a local or a global lowering of the design factors of safety. Non-structural cracking does not affect factors of safety. The following are some common causes of deterioration of concrete bridge structures:

(1) structural deficiency;
(2) corrosion of reinforcement;
(3) chemical attack;
(4) frost damage;
(5) internal reaction within the concrete;
(6) restrained movement;
(7) mechanical damage;
(8) cracks.

4.4.1 Structural Deficiency

Unfortunately, it is by no means uncommon for errors to be made in the design of all types of reinforced concrete structures (including bridges) or for mistakes to be made during site construction, causing a global lowering of the factors of safety.

These *structural* causes can only be overcome by strengthening the structures to reinstate the design factors of safety. Although most repairs are structural in nature (i.e. local lowering of factors of safety) they are generated from *non-structural* causes. All types of reinforced concrete structures, being very accommodating in nature, will relieve and redistribute load from any local area in distress.

4.4.2 Corrosion of Reinforcement

Corrosion is an electrochemical phenomenon; thus, in a strict sense, only conductive material can corrode—plastics and concrete deteriorate by chemical degradation. Corrosion begins with the natural tendency of steel, like all metals, to dissolve in water. Some metals such as steel and iron will go into solution much more readily than others such as gold or silver, which hardly revert at all. The metal, therefore, has an electrical potential. This does not, however, cause any problems when the environment immediately surrounding it has the same potential. With reinforced concrete, the steel is surrounded by highly alkaline concrete with pH 12·5–13·0, and this has a passivating effect on the surface of the steel which prevents corrosion from

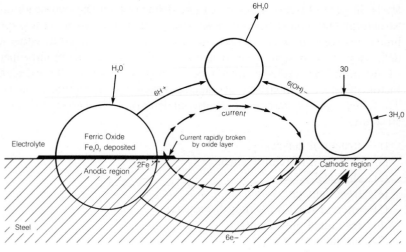

Fig. 4.6.   Passivating reaction (idealised).[6]

taking place (see Fig. 4.6). Almost invisible films of corrosion products adhere to the steel, stifling further corrosion and reducing corrosion reactions.

However, in the presence of acid gases or acid solutions the pH will drop as these substances react with the calcium hydroxide (which gives cement pastes their high alkalinity). As the pH falls to 9·5 or below, the passivating layer of ferric oxide becomes unstable and breaks down. A potential difference is set up between the steel and its environment (see Fig. 4.7), and an electrochemical reaction then starts, with electron flow from anodic to cathodic sites. Iron ions go into solution at the anodic site and react with water and oxygen (the fuels of corrosion) and precipitate as the loosely

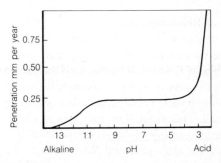

Fig. 4.7.   The effect of pH on the corrosion rate of mild steel.[7]

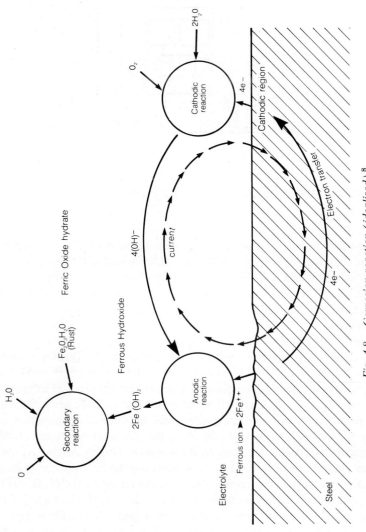

Fig. 4.8. Corrosion reaction (idealised).[8]

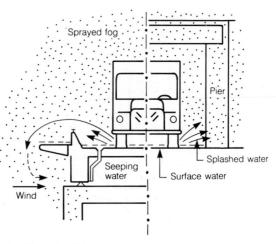

Fig. 4.9.   Chloride problems on a bridge deck due to vehicle spray.[11]

adhered ferric oxide hydrate (see Fig. 4.8). These oxides have a greater volume than the metal from which they were formed and thus overstress the surrounding concrete with subsequent spalling and rust staining. It can be seen, therefore, that it is of paramount importance to keep acid gases, acid solutions, water and oxygen out of the capillary and pore structure of the concrete.

Recent experience by the Department of Transport has shown that reinforcement in concrete bridge structures can, under certain circumstances, exhibit an unusual form of corrosion known as 'black rust'. For this type of corrosion to take place there must be an absence of oxygen at the anode. This usually means that the anode and cathode are located at some considerable distance from each other. In order to obtain a flow of electrons from anodic to cathodic sites, the electrical resistance of the concrete must be reduced, usually by the presence of highly conductive ions such as chloride or nitrate. This mechanism is different from the normal corrosion reaction as the reinforcement is converted to the black oxide Fe_3O_4 rather than dissolving to form ferric oxide or rust. The Fe_3O_4 is, in fact, a combination of ferrous and ferric oxide $(Fe_2O_3 . FeO)$ and is chemically similar to the mineral magnetite. It is a relatively unstable oxide and readily converts to ferric oxide when exposed to oxygen. This form of rust is very aggressive and, because it is not formed as a precipitate, no volume expansion occurs. There is, therefore, no visual warning of its occurrence. This is, potentially, a very dangerous form of corrosion and could lead to premature failure of a member or structure.[9]

Corrosion of reinforcing steel can result from four main causes:

(a) carbonation or sulphation;
(b) chloride attack;
(c) inadequate cover;
(d) cracks.

Because, in most cases, initial corrosion of steel is localised, significant changes in structural strength of the composite material result from the loss of quite small quantities of steel. Corrosion also tends to be progressive, since the bulky corrosion by-products overstress the concrete cover causing spalling and exposure of fresh steel to the corrosive medium.

4.4.2.1 Carbonation

All calcareous cements react with water in concrete mixes to produce various hydration products which provide a highly alkaline environment (pH values from 12 to 13) in wet or moist concrete. On exposure to the atmosphere, however, air permeates the concrete to some extent over a period of time. Although carbon dioxide gas is present in normal outdoor air at only a low concentration of about 0·3% by volume, it reacts with and decomposes the accessible hydrated cement compounds, causing a significant reduction in the alkalinity of the concrete. The process, advancing inwards from the exposed concrete surface, is referred to as carbonation. The outside layer of concrete in which appreciable reaction with carbon dioxide has taken place, giving reduced alkalinity, is termed the carbonated layer.

The principal carbonation reaction, which occurs initially, is the conversion of calcium hydroxide, by carbon dioxide gas dissolved in the pore water, into calcium carbonate and water. However, sodium and potassium hydroxides released from the cement can also react in a similar manner. At a later stage, as carbonation proceeds, the remaining cement hydration products, consisting of calcium silicate hydrates, aluminates and alumino-ferrites, or related complex hydrated salts, are attacked and decomposed. Calcium carbonate and hydrated silica, alumina, ferric oxide and calcium sulphate are ultimately formed, the latter being derived from the setting regulator added to Portland cement.

Carbonation takes place slowly over a period of years with well-compacted, dense concrete of low permeability (exposed outdoors) and is generally limited to a depth of several millimetres in the surface layer. The main body of the material thus remains unaffected.

Moisture is required for the carbonation reaction, which takes place (as described above) as a solution reaction. The amount of water condensed in the surface pore and capillary structure of the concrete has an influence on the rate at which carbon dioxide can penetrate and, therefore, the rate of carbonation. Carbonation occurs most rapidly at humidities in the range of 50–70%. Below 50% a moisture film does not tend to form, whilst above 75% the pores fill with water and become blocked. However, any cracks in the concrete provide a path for the ingress of air, which will allow carbonation to penetrate to greater depths. Other circumstances, particularly those associated with a reduction in concrete quality, i.e. weak, porous concrete with high permeability, may also result in the depth of the carbonated layer being considerably greater.

The principal carbonation reaction is

$$Ca(OH)_2 + CO_2 \xrightarrow[\text{'H}_2\text{O}]{} CaCO_3 + H_2O$$

Sulphation is a similar reaction to carbonation, where atmospheric sulphur dioxide causes loss of alkalinity of the concrete. The combination of the carbonation and sulphation leads to slightly greater rates of loss of alkalinity; therefore, increased cover to reinforcement is recommended in atmospherically polluted areas. It has been found that the depth of carbonation is approximately proportional to the square root of time:[10]

$$D = K\sqrt{t}$$

where D = depth of carbonation (mm), t = age of concrete (years) and K = a constant for the actual concrete. It follows that if the depth of carbonation at a point in a structure at a particular age is known and 'constant' conditions are assumed, then the depth of carbonation at a later age, or the further time taken to reach a given depth of carbonation, can be estimated. For example, if the depth of carbonation of a given concrete, D, is 20 mm after 16 years, the K value for the concrete can be calculated as follows:

$$20 = K\sqrt{16} \qquad K = 5$$

Therefore, the depth of carbonation after 25 years is given by

$$D = 5\sqrt{25} = 25 \, \text{mm}$$

Conversely, the time taken for the depth of carbonation to exceed the cover to the reinforcement of, say, 30 mm is given by

$$30 = 5\sqrt{t} \qquad t = 36 \, \text{years}$$

Assuming that the cover to the steel was as designed, 30 mm, the structure has a remaining predicted corrosion-free life of 36 years. Thus, for a 120-year design life, the anticipated actual life free from corrosion is only

$$36 \times \frac{100}{120} = 30\% \text{ of the design life}$$

4.4.2.2 Chlorides

Chloride ions are the most common materials which can break down the passive protective layer on steel in concrete. In carbonated concrete the depassivation takes place very quickly, with smaller amounts of chloride, and steel corrosion is accelerated.

Earlier work indicated that flake or solid calcium chloride ($CaCl_2$) added at or below 2% by weight would cause little harm to embedded metal. In practice, this dosage rate was difficult to adhere to when using hydrated flake calcium chloride, with resultant uneven distribution and large local concentrations. These figures are now known to be very high. The recently superseded Code of Practice CP 110 for normal reinforced concrete states that the chloride ion concentration should not exceed 0·35% by weight of cement for 95% of test results, with no result to exceed 0·5%. This equates to a maximum anhydrous calcium chloride content of half that which was earlier considered to be a safe level (see Table 4.3). Table 4.4 shows the limits of chloride content of concrete allowable in the British Standard BS 5400: Part 8: 1978.[16a]

Sodium chloride from washed de-icing salts on highway structures has now been found to present serious problems (see Fig. 4.9). It is not uncommon for the de-icing salts to be washed sideways over bridge deck

TABLE 4.3
Specified and Equivalent Maximum Allowable Amounts of Chloride Compounds and Ions in Concrete[12]

Reference	% $CaCl_2$ (w/w cement)	% $CaCl_2.2H_2O$ (w/w cement)	% Cl (w/w cement)	% Cl (w/w concrete)
13	1·5	2·0	0·96	0·12
14 95% results	0·55	0·73	0·35	0·04
All results	0·78	1·04	0·5	0·06
15 Low risk	0·63	0·83	0·4	0·05
Medium risk	1·56	2·08	1·0	0·15
High risk	>1·56	>2·08	>1·0	>0·15

TABLE 4.4
Limits of Chloride Content of Concrete Allowable in BS 5400: Part 8: 1978[16a]

Type or use of concrete	Maximum total chloride content (expressed as percentage of chloride ion by weight of cement)
Concrete for any use made with cement complying with BS 4027 or BS 4248	0·06
Reinforced concrete with cement complying with BS 12 Plain concrete made with cement complying with BS 12 and containing embedded metal	0·35 for 95% of test results with no result greater than 0·50

Note: % chloride ion × 1·648 = % equivalent sodium chloride.
　　　% chloride ion × 1·565 = % equivalent anhydrous calcium chloride.

parapets and blown onto the support structure, with subsequent absorption of up to 200 mm depth into the concrete.

Marine environments present similar problems. Chloride ions which have penetrated from the surface of the concrete, as in the latter two examples, present a more serious risk to corrosion than ions inherently present from aggregates or chemical admixtures on a weight for weight basis. This is because chlorides in the original mix tend to be partially chemically bound with hydration by-products, leaving less to migrate with the pore water. With porous or cracked concrete, or if the cover to the reinforcement is inadequate, the rate of corrosion of the steel under these conditions may be greatly enhanced.

4.4.2.3 Cover to Reinforcement

From Tables 4.5, 4.6 and 4.7 (which are extracts from BS 5400: Part 4: 1984 and BS 8110: Part 1: 1985,[16,17]) it can be seen that the designed depth of cover to the reinforcement can be based on several criteria:

(a)　the compressive strength of the concrete;
(b)　the condition of exposure;
(c)　water/cement ratio;
(d)　minimum cement content.

The argument for (a) is that the durability of the concrete increases with compressive strength (this is still a highly contentious issue) whilst the argument for (b) is that more concrete protection is required when the

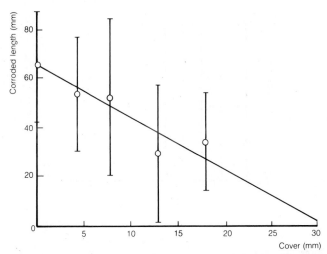

Fig. 4.10.   Influence of cover on corroded length.[18]

exposure conditions become more aggressive (i.e. water, chemicals, abrasion). Specification of the maximum free water/cement ratio is, though, the most important aspect.

Figure 4.10 highlights the importance of the concrete cover in protecting steel from corrosion, the corroded lengths shortening dramatically as the cover increases.

4.4.2.4 Cracks
Flexural cracks (inherent in the design of reinforced concrete) of the order of 0·2–0·5 mm in width are not thought to be large enough to act as prime corrosion initiators. In the absence of cyclic live load these can, in fact, autogenously re-seal.

Figure 4.11 indicates that the corroded lengths are constant when crack widths are within the range of 0·2–0·5 mm, thus showing that, in general, cover to concrete is more important than flexural cracking of this type.

Many other mechanisms can cause concrete to crack, resulting in crack widths significantly greater than 0·5 mm. These cracks can then initiate corrosion, as shown in Fig. 4.12, by allowing all the requirements for corrosion to be present. The inside of the crack is carbonated, lowering the pH and locally de-passivating the steel, allowing a galvanic cell to form. It is also wide enough to allow the fuels for corrosion to enter, i.e. water, oxygen and also, when present, chloride ions. Table 4.8 is an extract from BS 5400: Part 4: 1984, giving design values of crack width.[16]

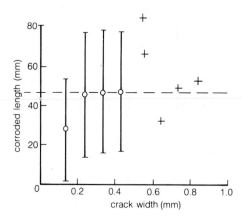

Fig. 4.11.   Influence of crack width on corroded length.[18]

4.4.3 Chemical Attack

Chemical attack occurs when aggressive chemical liquids come into contact with the concrete and soften or otherwise damage the surface. Also, aqueous solutions of sulphates disrupt the hardened cement in concrete by reacting with the calcium aluminate hydrate (C_3AH_6) to form expansive calcium sulpho-aluminates. This problem is usually associated with concrete in substructures and is of great importance when considering foundations in sulphate-bearing soils.

4.4.4 Frost Damage

One of the most common causes of spalling of concrete in an open environment (i.e. exposed slabs, deck) is attributed to freeze–thaw cycles.

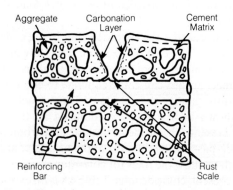

Fig. 4.12.   A large crack initiating corrosion.

TABLE 4.8
Design Crack Widths[16]

Environment	Examples	Design crack width (mm)
Extreme Concrete surfaces exposed to:		0·10
abrasive action by sea water	Marine structures	
or		
water with a pH ≤ 4·5	Parts of structure in contact with moorland water	
Very severe Concrete surface directly affected by:		0·15
de-icing salts	Walls and structure supports adjacent to the carriageway Parapet edge beams	
or		
sea water spray	Concrete adjacent to the sea	
Severe Concrete surfaces exposed to:		0·25
driving rain	Walls and structure supports remote from the carriageway	
or		
alternate wetting and drying	Bridge deck soffits Buried parts of structures	
Moderate Concrete surfaces above ground level and fully sheltered against all of the following:		0·25
rain de-icing salts sea water spray	Surface protected by bridge deck waterproofing or by permanent formwork Interior surface of pedestrian subways, voided super-structures or cellular abutments	
Concrete surfaces permanently saturated by water with pH > 4·5	Concrete permanently under water	

Porous or cracked concrete (which is not purposely air-entrained) will absorb water into the surface pores, capillaries and cracks during the wetting cycle; as the water freezes, it will expand 9% in volume. If the water is restrained from expanding, a pressure will build up and this will depress the freezing point. The pressure required to depress the freezing point is shown in Table 4.9. At the minimum temperature of $-20°C$ likely to be

TABLE 4.9
Pressure Required to Depress Freezing Point

Pressure on water (N/mm^2)	Freezing point (°C)
0	0
110	−10
190	−20

encountered in this country, ice generates a considerable pressure under restrained conditions. This will be higher than the resisting internal bonding structure of the concrete, and thus the induced shear stresses will spall the concrete surface. Unless the concrete is protected, this cycle will repeat itself until the concrete is completely destroyed.

4.4.5 Internal Reaction within the Concrete

Alkali–silica reaction (ASR) and alkali–carbonate reaction (ACR) are specific forms of alkali–aggregate reaction (AAR) which are internal reactions within the concrete. ASR is a mechanism of deterioration of concrete resulting from an interaction of hydroxyl ions in the pore water within the concrete and certain types of reactive silica occasionally found in the aggregate. The product of the reaction is a gel which imbibes water and alkalis from the pore fluid and produces a volume expansion. This increase in volume induces tensile stresses within the structure of the concrete which can ultimately be higher than the resisting stresses, with resulting internal fractures and surface cracking. Differential expansions within the structure of the concrete have the same net effect of overstressing. For ASR to take place, a combination of sufficient water, sufficient alkali and a critical amount of reactive silica in the aggregate are all required. Removal of any one can prevent the reaction from occurring. ACR follows the same requirements with the exception that the aggregate must contain a critical amount of reactive carbonate.

4.4.6 Restrained Movement

Concrete, in common with most other materials, changes volume with variation in temperature (either expansion or contraction). Also, as concrete begins to set and harden, it shrinks, because the cement paste in the matrix changes volume as the chemical reaction between the cement and the water takes place. This reaction generates heat known as heat of hydration. There is also a possibility of further shrinkage if water

TABLE 4.10
Types of Cracks[19]

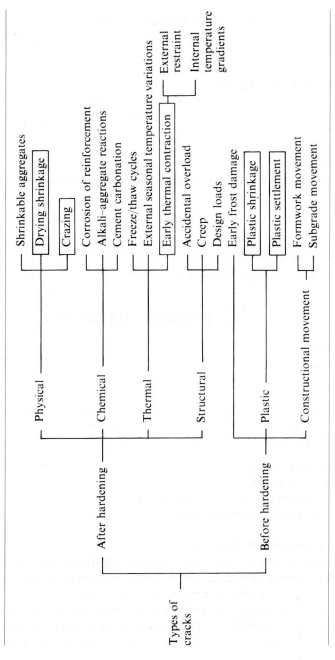

evaporates too quickly from the surface as the concrete dries out. If any change in volume is restrained in any way, then damage (spalling and cracking) may occur.

4.4.7 Mechanical Damage

It is not unusual for vehicles to hit highway structures and break away pieces of concrete. This may then expose the steel reinforcement to aggressive agencies which will lead to corrosion and further deterioration.

4.4.8 Cracks

Non-structural cracks can be considered under the following headings:

(a) plastic—which occur within the first few hours;
(b) thermal contraction—which occur within one day to two or three weeks;
(c) drying shrinkage—which occur after several weeks or even months.

Types of cracks are shown in Table 4.10.

4.5   MATERIALS FOR REPAIR

A basic first principle for the repair of reinforced concrete structures is that the repair medium should be as close as possible in all physical characteristics (Young's modulus, coefficient of expansion, strength) to the base concrete. This is aimed at ensuring that the properties of the old and new work are similar and so enhance the likelihood of achieving and maintaining a good bond by limiting the boundary stresses.

The materials for all types of repair fall in four general areas:

(1) cement-based;
(2) resin-based;
(3) polymer-modified cement-based;
(4) cement/pozzolanic-modified.

4.5.1 Cement-based Materials

A variety of cements, ordinary Portland, rapid-hardening Portland and sulphate-resisting Portland, can be used to produce grouts for crack filling and, when mixed with aggregates, produce a repair mortar.

When employing this method, it is important to ensure that the aggregates are as similar as possible to those employed in the base concrete and that the repair is carefully cured to prevent drying shrinkage. Both of these movements would induce stresses at the interface, with possible subsequent disbonding. It is, however, possible to enhance the performance of the repair material by using one of several chemical admixtures. These would most commonly be employed for the following reasons:

(a) to increase the density and/or workability (thus the waterproofing capabilities) of the mix;
(b) to accelerate the rate of hardening;
(c) to dramatically increase the workability (with a superplasticiser for thin-section major repair work).

4.5.2 Resin-based Materials

4.5.2.1 Polyester Resins

These are oil-derived resin systems usually applied in adhesive or mortar form. Polyester resins have good general adhesion and rapid curing on dry concrete, but perform poorly in damp or wet conditions. They are not normally applied in thick layers because of the possibility of shrinkage during curing. Unfortunately, the cure of polyester resin systems is exothermic and this can result in significant thermal movements, inducing high internal stresses.

4.5.2.2 Epoxy Resins

These, too, are oil-derived, with a history in the construction industry of some 30 years. Modern, well-formulated systems give excellent adhesion and performance in dry, damp and wet conditions. Although the nature of cure is exothermic, thermal shrinkage can be reduced to low levels.

Epoxy resins cover a whole range of materials, including low-viscosity grouts (very useful for grouting fine structural cracks), many mortars with different fillers or aggregates, and protective coatings. Unfortunately, epoxy resins have, in general, considerably higher coefficients of expansion and lower values of Young's modulus than concrete. These disparities can induce high shear stresses at the interface of the repair and the base concrete. If these shear stresses exceed the tensile resisting stress of the base concrete, then the repair will subsequently disbond. It should not, therefore, be assumed that, because of the greatly superior compressive and tensile strengths of epoxy resin mortars, a repair will give monolithic performance. The following performance properties of epoxy resins should be taken into consideration before use:

(1) high or very high compressive and tensile strengths;
(2) high adhesion to concrete and steel;

(3) good chemical and water-vapour resistance;
(4) lower Young's modulus than concrete;
(5) higher coefficient of thermal expansion than concrete;
(6) fast curing, even at low temperatures.

4.5.3 Polymer-modified Cement-based Materials

Polymer dispersions, as used in a cementitious environment, are a series of discrete 'plastic' spheres in a solution of water. The spheres can come in a variety of different degrees of 'hardness' and 'stickiness' which impart different properties to the modified cement paste. As these spheres can be very small, of the order of $0.1 \mu m$ in diameter, an anionic surfactant is required in the water suspension to stop the particles sticking together and forming large agglomerates. Once in the cement mix, the plastic spheres physically stick together to form thin thermoplastic films. It should be stated that very little chemical bonding takes place (coalescing) and that the films form physically at various temperatures (film-forming temperatures). As the particles are so small, a great deal of pore and capillary blocking can take place. On the basis of this understanding, the mechanisms by which polymer dispersions enhance the performance of cement pastes can be explained.

(a) *Increased adhesion*
     The introduction of 'sticky' plastic spheres makes the cement paste more adhesive and cohesive.
(b) *Increased flexural strength*
     Relatively soft but strong plastic films give a greater degree of elasticity but also act as reinforcement, giving increased flexural strength.
(c) *Increased abrasion resistance*
     A polymer-rich surface has essentially a plastic coating which is better at resisting abrasion than an unmodified cement paste.
(d) *Reduced shrinkage*
     The polymer spheres are small and can therefore block pores and capillaries and subsequently prevent water loss, reducing shrinkage.
(e) *Improved chemical resistance*
     The plastic coating on the surface has a much greater resistance to a variety of chemical attack than an unmodified cement paste.
(f) *Reduced permeability*
     A similar mechanism to (d); capillary pores are blocked, reducing permeability.

4.5.3.1 Types of Polymer Dispersion

Polyvinyl acetate (PVA)—this is a first-generation polymer and is only normally used as a bonding coat for various building materials. Not suitable for exterior use due to poor water resistance.

Styrene–butadiene rubber (SBR)—a second-generation synthetic polymer used as an admixture for grout, mortar and concrete to enhance the performance as follows:

(a) superior adhesion in dry, damp and wet conditions;
(b) higher flexural and tensile strengths;
(c) better abrasion resistance;
(d) reduced shrinkage;
(e) improved bond and chemical resistance.

Acrylic co-polymers—these have the following general properties:

(a) excellent adhesion in permanently dry conditions;
(b) mortars of high compressive strength may be produced;
(c) good abrasion and chemical resistance.

Styrene–acrylic co-polymers—this generation of polymers has been developed by chemically combining the SBR and acrylic bases to form a polymeric range of admixtures with the following properties:

(a) excellent adhesion in dry, damp or permanently wet conditions to concrete and steel;
(b) high compressive and tensile strengths;
(c) high abrasion resistance;
(d) high water impermeability;
(e) enhanced resistance to freeze/thaw cycles, alkalis and dilute acids;
(f) low shrinkage;
(g) high inherent alkalinity.

4.5.4 Cement/Pozzolanic Modified Materials

Portland cement blended with one of the following—

(a) micro-silica
(b) PFA
(c) GGBFC

—can dramatically reduce the permeability and hence the chloride ion diffusion of concrete.

Micro-silica, particularly, has the ability to block pores (mean particle

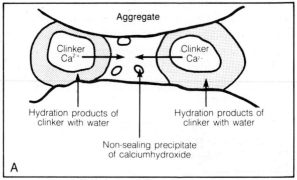

A) Hydration of Portland cement.

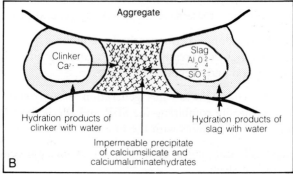

B) Hydration of concrete containing ggbfs.

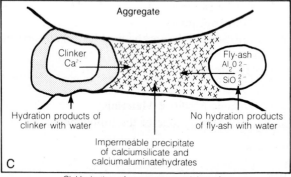

C) Hydration of concrete containing pfa.

Fig. 4.13.   Hydration of Portland cement and of Portland cement modified with pozzolanic materials.[20]

size 0·1 μm) by reacting with the dissolved calcium hydroxide liberated in the cement hydration process and acting as a superpozzolan to produce stable insoluble silicate and aluminium hydrates. Both PFA and GGBFC have a similar effect (see Fig. 4.13). This group, when combined with the benefits of polymer modification, can offer good engineering solutions.

A good repair material has the best combination of the following properties:

(1) mechanical properties as close to the base material as possible;
(2) good adhesion in dry, damp or wet conditions;
(3) low shrinkage (both during curing and long-term).

It is obviously not always possible to fulfil all of these criteria; therefore, there are differing situations which dictate selection of repair materials from each of the three basic groups.

4.6   CONCRETE REPAIR METHODS

Before any repair is considered a detailed diagnosis of the defects should be carried out. This aspect is dealt with in an earlier chapter in this book. Once the causes have been established and assessed, the objectives should then be decided upon; i.e. what are the repairs intended to achieve?

Demolition might be considered appropriate if, for example:

(a) the structure had become redundant (unlikely with a bridge);
(b) the structure had serious technical defects taking it beyond the realms of economic repair;
(c) financial constraints or political influences prevented action being taken.

On the assumption that repairs are to be carried out, the repair and protection system employed must:

(a) restore durability;
(b) restore structural integrity and act compositely with the rest of the structure;
(c) provide an aesthetically pleasing protective finish.

4.6.1 Consideration of the Repair Method

When considering methods for repair, the first principle should be that the repair medium should be as close as possible in all physical and chemical characteristics to the concrete to be repaired. The following must also be carefully examined:

(a) mechanical strength;
(b) protection of the reinforcement;
(c) adhesion between the repair and original concrete;
(d) bond between the repair and the steel reinforcement;
(e) shrinkage;
(f) thermal movement;
(g) durability;
(h) ease of application;
(i) cost.

Categories of concrete repair include:

(1) hand-applied mortars;
(2) sprayed concrete or mortar;
(3) repairs to cracks;
(4) recasting;
(5) surface coating.

4.6.2 Surface Preparation

Irrespective of the chosen repair method, good surface preparation is of paramount importance in order to achieve a homogeneous repair with good bond characteristics.

The first step is to mechanically cut away all spalled and loose concrete plus other identified areas of unsound concrete. Whenever possible, the concrete should be removed to expose the full circumference of the steel reinforcement and should extend along the length of the corroding bar for at least 50 mm beyond the point at which corrosion is visible. Extensive cutting-out may lead to the need for extensive propping. 'Feather edging' is not recommended and, wherever possible, the edges of the repairs should be sawn or disc-cut with power tools.

All exposed reinforcement should then be cleaned to remove scale and rust. The most effective method is grit blasting, preferably back to bright metal in accordance with the Swedish Standard SIS 05 5900-1967 SA2½.[21] Similarly, surrounding areas of defective concrete can be treated to remove loose, friable material. If environmental or economical constraints preclude the use of blast cleaning and chlorides are absent from the concrete, then hand-held power tools may be used. It must be pointed out, however, that this method and, to an even greater extent, hand cleaning will only remove loose rust and scale and may, in fact, 'polish' oxides on the surface. If the corrosion is substantial or so severe as to necessitate replacement of reinforcement, then a qualified engineer must assess the available steel areas and bond lengths required against the imposed loads and load distribution capabilities of the structure. For primary tension and compression steel, a 10% reduction in diameter is normally considered to be the limit for acceptance. It must be ensured that all areas prior to treatment are free from all unsound materials:

    (a)   dust;
    (b)   oil;
    (c)   grease;
    (d)   corrosion by-products;
    (e)   organic growth.

4.6.3 Hand-applied Repair Mortar System (Cement-based)

Following the surface preparation procedures, the next step is to protect the reinforcement. The material chosen should preferably be a strongly alkaline, dense, cement-based coating, possibly with the addition of a polymer dispersion and corrosion inhibitors. This type of treatment will accelerate the re-passivation of the steel surface and act as a barrier to the fuels of corrosion (water and oxygen) and migrating chloride ions, if they are present in the concrete. The use of chemical converters such as phosphoric acid or similar is definitely not recommended as splashing of these substances on to the concrete is likely to be very detrimental.

The broken-back surface of the concrete should then be wetted to achieve a saturated but surface-dry finish. If the substrate is particularly porous, it is recommended that a 'bonding bridge' be applied to the concrete surface. The expression 'bonding bridge' is something of a misnomer as it should have the consistency of a surface impregnant. It should consist of the same constituents as the repair mortar, generally polymer, cement, fine sands, very fine powders. The principle is that the bonding bridge is absorbed into the concrete surface which will not then draw fine materials from the repair mortar at the interface. The repair mortar then has the designed polymer/cement ratio and water/cement ratio at the interface of the repair. Thick slurry-type bonding bridges can, in fact, reduce bond, particularly if they are allowed to dry and then overcoated. The impregnant-type bonding bridge will also act (to a degree) as a barrier to any migrating chloride ions in the concrete.

Before either the steel reinforcement protector or the bonding bridge has cured, the repair mortars should be applied. These mortars should be as rich, dense and impermeable as possible (whilst complying with the shrinkage criteria), and normally take the form of a polymer-modified cementitious material. A placing-type repair should then be effected, compacting and building up in layers to the required thickness and finish. Whilst the repair site should not suffer from further damage, it is possible that moisture and oxygen already in the concrete (and a change in electrical potential) will enable other areas of corrosion and spalling to develop in the course of time.

As a final precaution, it is advantageous, both technically and aesthetically (patched concrete can look very unsightly), to coat the whole of the structure with a specialist protective membrane.

4.6.4 Sprayed Concrete

Definitions—'Sprayed concrete' is a mixture of cement, aggregate and water, which may include fibres and/or admixtures, projected at high velocity from a nozzle into place to produce a dense, homogeneous mass. 'Gunite' is a term used for sprayed concrete where the maximum aggregate size is less than 10 mm. 'Shotcrete' is a term used for sprayed concrete where the maximum aggregate size is 10 mm or greater.

Each are old, established repair systems, being particularly useful for applications involving extensive work. There are two basic methods of spraying concrete: the wet process and the more common dry process. As the name implies, in the wet process all the constituents are pre-mixed conventionally and then sprayed. In the dry process, the dry powders and water come into contact only at the nozzle head and, therefore, require a great deal of expertise from the nozzleman.

Sprayed concrete does not rely upon formwork or plastering-type techniques; therefore sharp corners and smooth surfaces do not naturally occur. Aesthetic finishes, including colour (sprayed concrete is not generally consistent in colour), can be achieved if economically viable. The performance of the material may be enhanced by the inclusion of a polymer admixture.

4.6.5 Repairs to Cracks

Fine cracks (down to approximately 15 μm in width), either structural or non-structural, some of which may be of full-element depth, are best repaired with low-viscosity epoxy grouts and injection techniques employing either gravity, pressure or vacuum methods.

The concept of epoxy resin injection for the repair of structural concrete elements has been thoroughly investigated in both the laboratory and on site, the results confirming this as an effective method of repair. A typical technique employed for more than a decade is to attach a series of injection/ bleed nipples to the line of the crack while the crack surface is sealed. Injection from the bottom upwards allows the group to force the air out of the bleed nipples in succession, thus filling the whole of the crack with epoxy grout and restoring the integrity of the unit.

Larger, non-structural cracks may be repaired by 'chasing out' and filling them with either a polymer–cement mortar or a flexible filling compound, depending upon whether the crack is 'live'.

4.6.6 Recasting

This method is usually reserved for large-scale reinstatements and has several inherent disadvantages:

(a) difficulties with colour;
(b) difficulties with surface texture;
(c) minimum application thickness of 75 mm;
(d) batching and quality control with relatively small quantities;
(e) compacting;
(f) bond.

The concrete should be designed and placed in accordance with good working practice, with particular attention being paid to compaction, curing and protection. The design of the shuttering and methods of placement also needs careful consideration, the most common being the 'letter box' or 'fixed pouch' type with high-mobility concrete. The high mobility of the concrete should be achieved by the use of a high-grade water-reducing admixture and not by a high water/cement ratio.

4.6.7 Surface Coatings and Sealers

The whole aspect of coating concrete has recently taken on a new significance following the realisation that modern concrete bridges are not as durable as was expected.

Coatings and sealers generally fall into the following groups:

(a) water-repellant surface impregnants;
(b) surface hardeners and pore blockers;
(c) cement-modified polymer coatings;
(d) specialist elastomeric polymer membranes.

4.6.7.1 Water-repellant Surface Impregnants
These include the following hydrophobic substances:

> silicone resins;
> polysiloxanes;
> silanes;
> stearates.

The principle of hydrophobic impregnates is well known. After impregnation with one of the substances, the carrier fluid evaporates, leaving behind a hydrophobic layer in the pores and capillaries which exerts a water-repelling action. Certain substances, such as silanes, are more efficient, having a chemical bond with the cement paste.

There is little doubt that this group of materials forms an efficient hydrophobic barrier on the concrete surface, repelling water and (where external chlorides are present) stopping the ingress of chloride ions. There are, however, several points worthy of further discussion. It is difficult to ascertain to what depth the materials are absorbed in the surface capillary structure of the concrete and to check their working efficiency. Certainly, silicones, whilst having a good water-repelling action, react with and are degraded by the saturated calcium hydroxide solution in the pore and capillary structure. By repelling water from the surface structure of the concrete, access for the ingress of carbon dioxide gas is made easier (as none of the pores are filled with water) and the rate of carbonation of the concrete increases. As all reinforced concrete cracks, problems may occur if a crack forms after impregnation, thus forming a discontinuity in the hydrophobic layer.

4.6.7.2 Surface Hardeners and Pore Blockers
These are materials based upon calcium, sodium and potassium silicates. When used in a solution as a surface wash, these materials react with the calcium hydroxide created as a hydration by-product of Portland cement, and form insoluble stable silicate hydrates which increase the density of the surface structure of the concrete. This therefore has the effect of slowing the penetration of water and the ingress of chloride ions from external sources. The silicate hydrates formed are, however, porous when compared to the crystalline calcium hydroxide consumed in the reaction. This allows an easier passage into the concrete for acid gases, resulting in an increase in the rate of carbonation.

4.6.7.3 Cement-modified Polymer Coatings

When the relative humidity of the surface pore structure of concrete is above 75%, conventional water-based polymer coatings will not adhere. If the RH rises to 90%, even solvent-based coatings will not adhere. The necessity to protect wet concrete does, however, arise, as is the case with bridge piers in polluted or acidic waters. Cement-modified polymer coatings (where the polymer solids content is greater than 25% by weight of cement) and polymer-modified cement coatings have to be applied to saturated but surface-dry concrete. A variety of different polymer dispersions may be used, depending upon the flexibility required. Usually the cement gel structure is modified by the use of micro-silica in order to achieve a dense impermeable coating.

4.6.7.4 Specialist Elastomeric Polymer Coatings

A specialist membrane coating should have the following technical features in order to protect concrete:

(1) It should be permanently elastomeric in order to bridge cracks and compensate for the normal movements of the structure.

(2) It should have a high diffusion resistance to carbon dioxide and sulphur dioxide gases; thus the membrane will act as a barrier to slow down or prevent further carbonation or sulphation of concrete.

(3) It should be weatherproof to stop driven water entering the concrete and providing more free oxygen for corrosion reaction.

(4) Its permeability to water vapour should be such as to allow the free transfer of water vapour out of the structure.

(5) It should resist degradation by ultraviolet light.

(6) It should be abrasion resistant.

(7) In the event of any future maintenance being necessary, it should be easy to refurbish.

(8) As a secondary consideration, it should have aesthetic value, i.e. remain clean with low dirt retention and be attractive in both texture and colour.

(9) It should have a high resistance to the diffusion of chloride ions from external sources.

Coatings complying with the above features will protect concrete against all aspects of concrete deterioration caused by ingress of water, acid gases or chlorides, and will also enhance the aesthetics.

4.7 CONCLUSIONS

Considering the number of mechanisms discussed earlier by which reinforced concrete can deteriorate, it is statistically probable that a very high percentage of all reinforced concrete bridges will suffer some form of continuous distress during their life. Aggressive natural environments, industrial, atmospheric and ground pollution, poor detail design, poor construction and workmanship all contribute towards a structure's deterioration. Modern repair materials can now offer an excellent solution for most repair and maintenance problems.

It is, though, considerably more economical to protect a structure by regular inspection and maintenance carried out by qualified personnel than to await advanced costly deterioration before taking any action. Indeed, there is now very strong commercial and technical evidence to suggest that it is advantageous to coat all concrete bridge structures with an elastomeric membrane in order to protect a costly investment. Repairs to concrete bridge structures can be successfully accomplished, but only if meticulous care is taken at all stages. Experienced supervision is also essential, as is a specialist applicator with the necessary expertise. Only then can the bridge be expected to have a useful, serviceable life.

REFERENCES

1. Bakker, R. F. M., Illustration of permeability and porosity. Diffusion within and into concrete. Paper presented at *13th Annual Convention of the Institute of Concrete Technology, University of Loughborough, 25–27 March 1985*, Institute of Concrete Technology, Slough, 1985.
2. Lydon, F. D., *Concrete Mix Design*, Applied Science Publishers, 1972.
3. Verbeck, G. J. and Helmuth, R. H., *Proc. 5th International Symposium on Chemistry of Cement, Tokyo, 1968*, Vol. 3.
4. Powers, T. C., Copeland, L. E., Hayes, J. C. and Mann, H. M., Permeability of Portland cement paste, *ACI J., Proc.* Vol. 51.
5. Powers, T. C., *Proc. 4th International Symposium on Chemistry of Cement*, Washington, 1960.
6. Beeby, A. W., *Concrete in the Oceans*, Technical Report No. 1, Cracking and Corrosion, CIRIA/UEG, Cement and Concrete Association and Department of Energy, 1978.
7. King, R. A., *Corrosion in steel reinforced concrete*, Corrosion and Protection Centre, Industrial Services, UMIST, Manchester.
8. Beeby, A. W., *Concrete in the Oceans* (see ref. 6).
9. Lloyd, C. G., Private communication, Liquid Plastics Ltd, Preston.
10. Tuuti, K., *Corrosion of Steel in Concrete*, Stockholm, CBI, 1982.

11. Browne, R. D., Practical considerations in producing durable concrete, *Improvement of concrete durability*, Thomas Telford, 1985.
12. Liquid Plastics Ltd, Internal publication.
13. Concrete Society, *Admixtures for Concrete*, Technical Report TRCS 1, 1968.
14. Building Research Establishment, *The Durability of Steel in Concrete*, Digest 264, July 1982.
15. Concrete Society, *Repair of concrete damaged by reinforcement corrosion*, Technical Report No. 26, October 1984.
16. British Standards Institution, BS 5400: Part 4: 1984, *Code of Practice for design of concrete bridges*.
16a. British Standards Institution, BS 5400: Part 8: 1978, *Recommendations for materials and workmanship: Concrete reinforcement and prestressing tendons*.
17. British Standards Institution, BS 8110: Part 1: 1985, *Structural use of concrete. Code of Practice for design and construction*.
18. Beeby, A. W., Carpentier, L. and Sorety, S., Cracking and design in reinforced concrete, *Proc. Symposium on Corrosion of Steel Reinforcement in Concrete Construction*, Society of Chemical Industry, 1979. Also: Contribution à l'étude de la corrosion des armatures dans le béton armé, *Annales de l'Institut Technique du Bâtiment et des Travaux Publics*, 1966.
19. Concrete Society, *Non-structural cracks in concrete*, Technical Report No. 22.
20. Higgins, D., Reducing the ingress of de-icing salts into concrete, *Construction Repairs and Maintenance*, 1986.
21. Swedish Standards Institution, Swedish Standard SIS 05 5900-1967 SA2$\frac{1}{2}$, *Pictorial surface preparation standards for painting steel surfaces*.

Note: Extracts from British Standards are reproduced by permission of the British Standards Institution. Complete copies can be obtained from the BSI at Linford Wood, Milton Keynes MK14 6LE, UK.

5

The Thermal Response of Concrete Bridges

M. J. N. Priestley

*Department of Applied Mechanics and Engineering Sciences,
University of California (La Jolla), San Diego, California, USA*

5.1 INTRODUCTION

The design of concrete bridges for temperature fluctuations has traditionally been a simple matter. Longitudinal movements induced by the maximum expected average temperature changes, typically $\pm 20°C$, are accommodated by the provision of sliding joints, bearing displacements, or by a flexible pier design. It is, however, only comparatively recently that thermally-induced stresses in bridge structures have received much attention. In part, this results from the introduction of stiff, efficient prestressed sections that have been developed over the past 20 years, which are more sensitive to strain-induced stresses, of which temperature stresses are an example. Reports of distress induced in European prestressed bridges by temperature gradients[1,2] began to surface in the 1960s and 1970s. In New Zealand, interest was stimulated in the late 1960s by severe cracking of Auckland's New Market Viaduct, a major urban prestressed concrete box girder motorway bridge. Measurements of the crack widths indicated a strong correlation with ambient temperature and solar radiation.

The problem results primarily from diurnal ambient thermal heating of continuous bridges. Under midsummer conditions, when high solar radiation levels which may exceed $1\cdot1\,k\,W/m^2$ on the deck surface at midday occur, the temperature of the deck surface rises to as much as $30°C$ above soffit temperature. This induces a tendency for the bridge to hog upward off internal supports. Dead weight or connection to the supports will restrain this, inducing sagging bending moments with tension stresses of significant magnitude developing at the soffit.

The temperature distributions induced are non-linear, causing local internal stresses which may add to the above effects, and which reduce shear strength.

A second thermal loading source is of significance to concrete bridge design. The heat released by hydration of cement during setting and curing of concrete is important for large sections whose ratio of volume to surface area results in low heat dissipation from the surfaces and a consequent temperature build-up in the concrete. Non-uniform temperature distribution during heat build-up and subsequent cooling can result in concrete cracking and overstressing of reinforcement. Although this is primarily a problem for massive structures such as dams and foundation pads, stresses induced in large bridge girders and diaphragms can be significant.

Complete thermal analysis of concrete structures is extremely complex. Stresses induced by temperature changes are influenced by creep and by the mechanical material properties (modulus of elasticity, tensile strength and coefficient of thermal expansion) which may be dependent on the effective

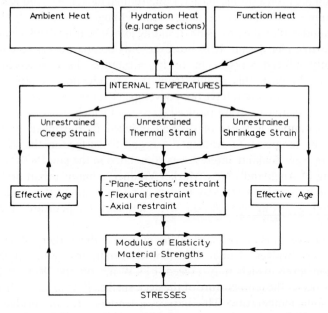

Fig. 5.1. Influence of thermal load on stresses in concrete bridges.

age of the concrete. The mechanical material properties, the creep and shrinkage characteristics and the effective age of the concrete are in turn dependent on the temperature/time history.

Figure 5.1 illustrates the interdependence of the main variables involved on the analysis path from heat input to thermal stresses, and includes three categories of heat source: ambient heat; hydration heat as discussed above; and function heat, where the structure is thermally loaded as a consequence of its primary structural role of containing hot or cold liquid or gas. The third category has some significance to response of concrete aqueducts, where the temperature of the contained water will modify ambient thermal response.

5.2 AMBIENT THERMAL LOADING

When the dominant thermal effect results from ambient heating, the interactions represented by Fig. 5.1 can be greatly simplified, and treated as a series of sequential and independent steps following the central path in the figure.

Ambient thermal response of bridge decks is a transient phenomenon influenced by many parameters, as indicated for a typical box-girder bridge in Fig. 5.2. In addition to the fundamental influence of the time-dependent

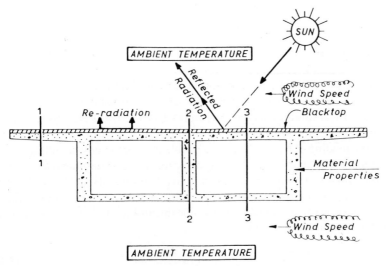

Fig. 5.2. Factors affecting ambient thermal response of a concrete bridge.

solar radiation, response is affected by ambient temperature and wind speed fluctuations, material properties, surface characteristics and section geometry. The design problem can be separated into three major phases. First, prediction of the critical design temperature distribution based on known local ambient characteristics (variation of solar radiation, ambient temperature, wind speed). Second, calculating stress levels induced in the bridge superstructure by the design thermal distribution. Third, assessment of the significance of the thermally-induced stresses to serviceability and ultimate load characteristics. The deformation nature of the imposed thermal loading makes this final phase of great importance.

5.2.1 Prediction of Temperature

The thermal response of an isotropic solid with a boundary, in contact with air, subjected to variations in ambient temperature θ_a and radiation q, as indicated in Fig. 5.3, is governed by the general Fourier conduction equation:

$$\frac{k}{\rho c}\left[\frac{\partial^2 \theta}{\partial x^2} + \frac{\partial^2 \theta}{\partial y^2} + \frac{\partial^2 \theta}{\partial z^2}\right] = \frac{\partial \theta}{\partial t} \qquad (5.1)$$

with the boundary condition

$$k\frac{\partial \theta}{\partial n} + q + h(\theta_a - \theta_b) = 0 \qquad (5.2)$$

where k is the conductivity, ρ the density and c the specific heat of the solid; θ_b is the solid boundary temperature, h the boundary heat transfer coefficient (a variable mainly dependent on the speed of air across the boundary) and n the direction normal to the boundary.

Since q, θ_a and h will in general be complex functions of time, direct mathematical solution of eqns (5.1) and (5.2) is impossible, and use of three-dimensional finite element techniques is appropriate. However, for ambient thermal loading of bridge decks, the problem can generally be substantially simplified. Thermal variation in the direction of the longitudinal axis of the bridge will not normally be significant, so a two-dimensional finite element representation of the bridge section may be analysed. For most bridge sections, transverse heat flow is insignificant, and eqn (5.1) can be simplified to the one-dimensional form:

$$\frac{k}{\rho c}\frac{\partial^2 \theta}{\partial y^2} = \frac{\partial \theta}{\partial t} \qquad (5.3)$$

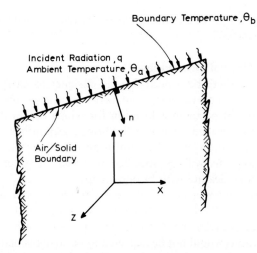

Fig. 5.3.   Coordinate definition for transient heat-flow analysis.

Equation (5.3) may be solved, in conjunction with eqn (5.2), by finite difference techniques. Agreement between finite difference solutions and both two-dimensional finite element and experimental solutions has been very close. Figure 5.4 shows a typical comparison between measured web temperatures and predictions of a two-dimensional finite element analysis and a linear heat-flow analysis, for thermal loading of a prestressed box-girder bridge.[3] The agreement of both methods with measured temperatures is good, and there is no noticeable reduction in accuracy with the simpler linear heat-flow model.

To analyse the thermal response of a complex section, such as that represented by Fig. 5.2, by finite difference techniques, it may be necessary to analyse more than one vertical line and combine the total results for the

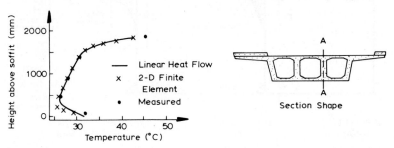

Fig. 5.4.   Theoretical and measured web temperatures for a box-girder bridge.

section. In Fig. 5.2, three characteristic areas are identified for analyses by lines 1-1, 2-2 and 3-3.

5.2.2 Design Thermal Distributions

Only rarely will there be a need for the designer to carry out thermal analyses to predict the level of thermal distribution appropriate for design. Such a need might occur in a location for which design data have not previously been specified, or where the bridge section is so unusual that established distributions are of doubtful applicability. Generally, design thermal distributions will be specified in codes of practice or design recommendations dealing with bridge loading.

Examples are shown in Fig. 5.5 of thermal gradients commonly specified for bridge design in the USA,[4] the United Kingdom[5] and New Zealand.[6] The differences between the distributions are rather surprising, as peak thermal conditions would not be expected to be so dependent on location. Particularly surprising is the marked difference between the distribution proposed by the American Prestressed Concrete Institute (Fig. 5.5a) for the USA and the New Zealand distribution (Fig. 5.5c) as both countries span similar ranges of latitude, and would therefore presumably be subject to similar peak solar radiation intensities.

The New Zealand design distribution was developed from a computer parameter study of the influence of wind, ambient temperature variation, blacktop thickness and surface solar absorptivity on the thermal response of a number of typical prestressed concrete bridge sections, including slabs, T-beams and box girders.[3] Critical conditions were found to occur on days

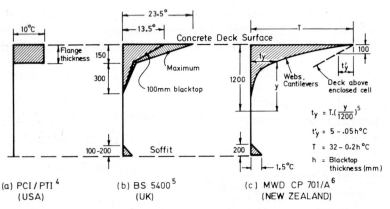

Fig. 5.5.   Design temperature distributions for concrete bridges.

of high solar radiation and low wind speed. Overlaying the concrete deck with a bituminous wearing layer (blacktop) resulted in a thermal insulating effect—a proportion of the daily solar energy input was required to increase the temperature level of the blacktop before the bridge was affected. Consequently, an increase in blacktop thickness can be expected to reduce the thermal effect.

The analyses indicated that the critical thermal conditions could be represented by the simple design distribution of temperature change shown in Fig. 5.5c. The distribution has two main components: fifth power decrease of temperature from maximum T (the value of which depends on the blacktop thickness) at the concrete deck surface to zero at a depth of 1200 mm, and a linear increase in temperature over the bottom 200 mm of the section. For superstructure depths less than 1400 mm these two components are superimposed. For concrete deck slabs above enclosed air cells, as is the case in box girders, a different distribution, shown dashed in Fig. 5.5c, is specified to represent the insulating effect of the air cell. This design distribution is currently adopted for all major concrete bridge design in New Zealand.

Thermal analyses indicate that design distributions are not likely to greatly exceed those of Fig. 5.5 for any latitudes between 45° S and 45° N, as peak solar radiation levels on horizontal surfaces between these latitudes do not vary significantly. At greater latitudes than 45°S or 45°N, reductions in solar radiation due to the angle of incidence will become increasingly significant. The contention that the design distribution of Fig. 5.5c is appropriate for other than New Zealand conditions is supported by recent measurements by Hoffman et al. of a full scale bridge in Pennsylvania.[7] It was found that peak design distributions agreed well in shape and magnitude with the New Zealand design gradient.

The design distribution of Fig. 5.5c is based on the premise that the critical thermal effect for concrete bridges is the thermally-induced tension stress on the soffit of continuous bridges. The critical distribution is thus the thermal profile maximising soffit tension stress, and not necessarily the profile with maximum temperature differential through the section. Analyses forming the basis for Fig. 5.5c were thus based on heat-flow analyses, as developed in the previous section, and simultaneous thermal stress analyses, using methods developed in subsequent sections of this chapter.

These analyses also enabled the influence of wind, ambient temperature variation, blacktop thickness and surface solar absorptivity on thermal response to be investigated. Figure 5.6 shows typical results for the interior

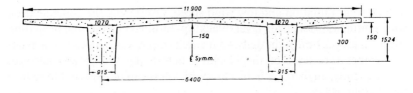

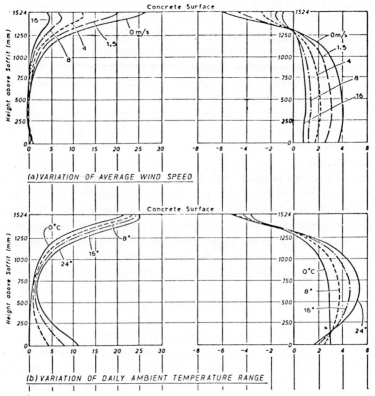

Fig. 5.6. *Influence of variables on temperature and thermal stress of a double-T continuous bridge.*

span of a continuous double-T section. It will be seen that an increase in average wind speed from 0 to 16 m/s (Fig. 5.6a) markedly decreases the design thermal gradient and stress levels. The influence of increasing the range of daily ambient temperature fluctuation (Fig. 5.6b) is more complex, increasing tension stress levels at heights above the soffit but decreasing the

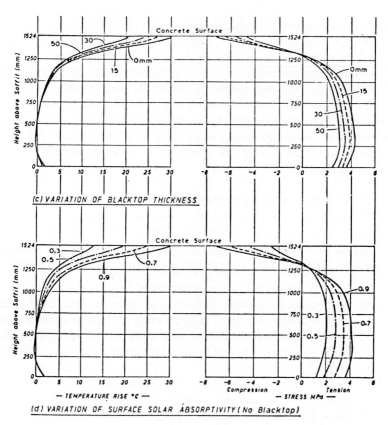

(c) VARIATION OF BLACKTOP THICKNESS

(d) VARIATION OF SURFACE SOLAR ABSORPTIVITY (No Blacktop)

— TEMPERATURE RISE °C —

Compression   Tension

— STRESS MPa —

Fig. 5.6—contd.

soffit tension stress. As noted earlier, increasing the thickness of the blacktop layer decreases thermal response. Figure 5.6c indicates a 30% reduction in temperature and stress resulting from overlaying a bare concrete deck with a 50 mm thick blacktop layer. Surface solar absorptivity (Fig. 5.6d), relating the amount of solar energy absorbed to the total normal incident radiation, is effectively a measure of the surface colour, and is significant for unsurfaced bridge decks. Light grey fresh concrete has an absorptivity of about 0·6 and, as Fig. 5.6d indicates, will be less subject to thermal stress problems than older concrete of higher absorptivity resulting from discolouration by tyres, engine oil and normal ageing processes.

The behavioural trends indicated by Fig. 5.6 are characteristic of all sections analysed and could be used to modify the profiles of Fig. 5.5c, where critical thermal distributions are known to vary from the base values of average wind speed $= 1\cdot5$ m/s, daily ambient temperature range $= 10°$C and surface absorptivity $= 0\cdot9$ adopted in preparing Fig. 5.5c.

In some locations, reversed thermal gradients, where the bridge deck is colder than the soffit, may also be significant. Reversed gradients can occur during summer conditions on cold, clear nights where heat is radiated from the deck surface to the night sky, or during winter, when snow or hail is deposited on the bridge surface while the bulk of the cross-section is at a significantly higher temperature. It will be rare, however, for such gradients to exceed about 10°C.

5.2.3 Stress Analysis of Uncracked Sections

With the temperature distribution specified, as a result of code requirements or a detailed heat-flow analysis, temperature stresses can be calculated by straightforward application of the principles of structural mechanics. The following assumptions are made in developing the theory:

(i) Material properties are independent of temperature.

(ii) Homogeneous isotropic behaviour is assumed. Thus temperature stresses can be considered independently of stress or strain imposed by other loading conditions, and the principle of superposition holds. Although this assumption is reasonable for fully prestressed bridges, modifications are necessary to cope with the influence of cracking with conventionally reinforced or partially prestressed structures. These are discussed in later sections of this chapter.

(iii) Initially plane sections remain plane after thermal loading.

(iv) A temperature distribution can be defined, throughout the structure, at which the structure is thermally stress-free.

Temperature stresses are induced not directly by the temperature changes themselves but by restraint of the thermally-induced expansions and contractions. The principle of thermally-induced stress is illustrated in its simplest form by Fig. 5.7, where a prismatic beam, length L, fixed at each end to rigid abutments, is subjected to a uniform temperature increase θ. If the beam were free to expand, the increase in length due to θ would be

$$\Delta L = \alpha\theta L \qquad (5.4)$$

where α is the linear coefficient of thermal expansion. If this deformation were free to occur, there would be no thermally-induced stresses. However, since this expansion is restrained by the rigid abutments, compression forces C must be developed such that an equal and opposite contraction ΔL

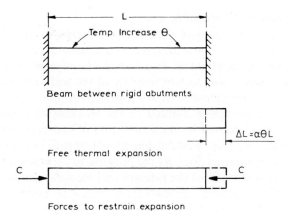

Fig. 5.7.   Uniform temperature rise of prismatic beam.

is induced. Thus, removing the incompatible expansion ΔL implies a compressive strain of

$$\varepsilon_c = \frac{\Delta L}{L} = \alpha\theta \tag{5.5}$$

and therefore a compressive stress of

$$f_c = E\alpha\theta \tag{5.6}$$

If the section area is A, the restraint force C is given by

$$C = EA\alpha\theta \tag{5.7}$$

Note that restraint of a temperature increase induces compressive stress, and conversely, restraint of a temperature decrease induces tensile stress.

Initially, two additional assumptions will be made to facilitate analysis, and the effects of these assumptions examined later.

(v) Temperature varies with depth, but is constant at all points of equal elevation.

(vi) Longitudinal and transverse thermal response of bridge superstructures can be considered independently, and the results superimposed. That is, the longitudinal and transverse thermal stress fields are supposed each to be uniaxial, without interaction. This simplifies analysis, particularly for complex section shapes, such as box girders, without inducing excessive errors.

5.2.3.1 Longitudinal Thermal Stress

It is convenient to separate longitudinal thermal response into two components and superimpose the results. First, the structure to be analysed is made statically determinate by removal of sufficient internal redundancies, and stresses due to non-linearity of the temperature profile are calculated. These are generally known as *primary temperature stresses*. The inadmissible deformations induced at the locations of released redundancies are then removed by the application of appropriate forces. The stresses induced by this re-establishment of compatibility are known as *secondary temperature* stresses. By the principle of superposition, the total state of stress is the sum of the primary and secondary stress sets.

Figure 5.8 illustrates this process for a three-span bridge, where solar heating of the deck creates a tendency for the superstructure to hog. Two possible methods of redundancy release are shown. In Fig. 5.8a vertical deflections at the internal supports are released, allowing the bridge to span between abutments. Primary thermal stresses and deflections are then calculated. Redundant vertical displacements induced by the primary stress field are removed by the imposition of appropriate vertical compatibility reactions P at the internal supports, and the secondary thermal stresses corresponding to the resulting induced moments calculated. An alternative and more useful technique, illustrated in Fig. 5.8b, releases rotations at the ends of each span. Continuity moments M are imposed to remove the distortion rotations ϕ induced at each internal support by the primary strain field.

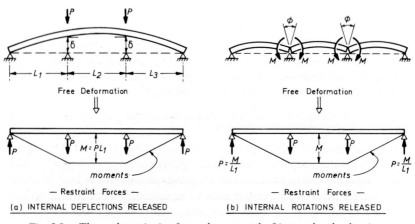

Fig. 5.8.   *Thermal continuity forces by removal of internal redundancies.*

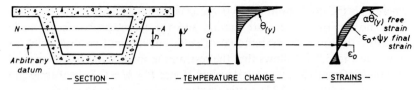

Fig. 5.9. Primary thermal strains.

Primary temperature stresses: Consider the general section in Fig. 5.9, subjected to temperature change from a thermally stress-free reference temperature profile. Although the section shown represents a box girder, the method is general, and applicable to all section shapes. Initially, it will be assumed that all points of constant elevation y have the same temperature change $\theta_{(y)}$. For totally unrestrained expansion at all elevations, the free strain profile would be

$$\varepsilon_f = \alpha \cdot \theta_{(y)} \tag{5.8}$$

where α is the linear coefficient of thermal expansion of the concrete. Since initially plane sections remain plane, the final strain profile $\varepsilon_{(y)}$, shown in Fig. 5.9, must be linear. The difference between the unrestrained strain $\alpha \cdot \theta_{(y)}$ and the final strain $\varepsilon_{(y)}$, shown shaded in Fig. 5.9, implies restraint stresses equal to

$$f_{p(y)} = E_c(\varepsilon_{(y)} - \alpha \cdot \theta_{(y)}) \tag{5.9}$$

where E_c is the concrete modulus of elasticity.

Integration of eqn (5.9) over the section depth d yields the axial force

$$F = E_c \int (\varepsilon_{(y)} - \alpha \cdot \theta_{(y)}) b_{(y)} \, dy \tag{5.10}$$

where $b_{(y)}$ is the net section width at height y.

Similarly, taking moments of the primary stress distribution $f_{p(y)}$ about the neutral axis NA at elevation $y = n$, the internal moment induced by $\theta_{(y)}$ will be

$$M = E_c \int (\varepsilon_{(y)} - \alpha\theta_{(y)}) b_{(y)} (y - n) \, dy \tag{5.11}$$

Now, as shown in Fig. 5.9, the linear final strain distribution $\varepsilon_{(y)}$ may be expressed as

$$\varepsilon_{(y)} = \varepsilon_0 + \psi \cdot y \tag{5.12}$$

where ε_0 is the final strain at $y = 0$ and ψ is the final curvature of bending. Since no external forces are acting, and the structure has been made statically determinate by removal of sufficient internal redundancies, as shown in Fig. 5.8, it follows that the restraint moments and forces implied by eqns (5.10) and (5.11) cannot develop. Hence $P = M = 0$. Substituting eqn (5.12) into (5.10) and (5.11), setting both to zero, and noting that

$$\int y \cdot b_{(y)} \, dy = n \cdot A \tag{5.13}$$

and

$$\int y^2 \cdot b_{(y)} \, dy = I + n^2 A \tag{5.14}$$

yields the following equations:

$$\psi = \frac{\alpha}{I} \int \theta_{(y)} b_{(y)} (y - n) \, dy \tag{5.15}$$

and

$$\varepsilon_0 = \frac{\alpha}{A} \int \theta_{(y)} b_{(y)} \, dy - n\psi \tag{5.16}$$

where A is the section area and I is the second moment of area of the section about the neutral axis. The primary stresses given by eqn (5.9) can be calculated as

$$f_{p(y)} = E_c(\varepsilon_0 + \psi \cdot y - \alpha \cdot \theta_{(y)}) \tag{5.17}$$

Secondary temperature stresses: For continuous beams, the curvature resulting from the primary thermal response is partially restrained by internal supports, inducing secondary thermal stresses. Figure 5.10 illustrates the analysis technique for calculating continuity moments resulting from this restraint for a three-span beam, using the method of rotation redundancies outlined in Fig. 5.8b. First, the beam is 'cut' at each internal support and allowed to hog (or sag) under the thermal curvature ψ calculated from eqn (5.15). Restraint moments M are then applied to each end of each span to remove the incompatible rotations. For prismatic sections, fixed end moments

$$M = - E_c I \psi \tag{5.18}$$

are required for each span.

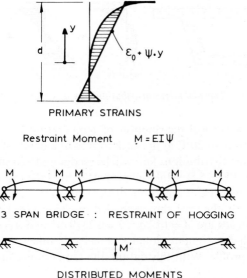

PRIMARY STRAINS

Restraint Moment $M = EI\psi$

3 SPAN BRIDGE :   RESTRAINT OF HOGGING

DISTRIBUTED MOMENTS

$M' \cong 1.17M$

Fig. 5.10.   Continuity moments from thermal loads.

The final stage in analysis is to release the end restraint moments, if appropriate, and calculate the final moments M' by moment distribution, or some such means, to obtain a moment profile of the form shown at the bottom of Fig. 5.10. The stresses induced by M' will be

$$f_{s(y)} = \frac{M'(y - n)}{I}$$

(5.19)

where y is still measured from the arbitrary datum as shown in Fig. 5.9. The total temperature stresses due to the vertical temperature gradient are then the combined stresses due to non-linearity of the temperature distribution, from eqn (5.17), and continuity, from eqn (5.19). That is,

$$f_{t(y)} = E_c(\varepsilon_0 + \psi y - \alpha\theta_{(y)}) + \frac{M'(y - n)}{I}$$

(5.20)

Figure 5.11 illustrates this process, and shows the typical shape of the primary, secondary and total thermal stress distributions.

A typical example of longitudinal thermal response is summarised in Fig. 5.12. The three-span prestressed concrete beam-and-slab bridge is

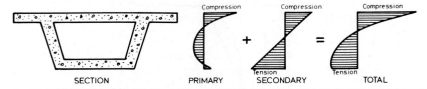

Fig. 5.11. *Thermal stress distributions for a continuous box girder.*

subjected to a vertical distribution of temperature increase, decreasing from a maximum of 30°C at the deck to zero at a depth of 1·2 m according to a fifth-power distribution. This will be recognised as a simplified version of the profile of Fig. 5.5c. Material properties are taken as $\alpha = 10^{-5}/°C$ and $E_c = 30$ GPa. Because of repetition of elements in the section, it is sufficient to consider only the area shown shaded in Fig. 5.12b. For this area, the section properties are $A = 1·12$ m^2, $I = 0·2319$ m^4 and $n = 0·752$ above the datum, which for computational simplicity is taken at the base of the temperature profile. With this datum, the temperature increase is

$$\text{for } y \leq 0 \qquad \theta_{(y)} = 0$$

$$\text{for } 0 < y \leq 1·2 \text{ m} \qquad \theta_{(y)} = \frac{30y^5}{1·2^5} = 12·06y^5 \, °C$$

Substituting this temperature distribution, and the section properties above, into eqns (5.15) and (5.16) yields the curvature ψ and strain at $y = 0$, ε_0, as

$$\psi = 19·1 \times 10^{-5} \text{ m}^{-1} \qquad \varepsilon_0 = -2·95 \times 10^{-5} \text{ m/m}$$

Thus the primary thermal stresses are

$$f_{p(y)} = 0·3(-2·95 + 19·1y - \theta_{(y)}) \text{ MPa}$$

The moment to restrain the thermal curvature will be

$$M = E_c I \psi = 30 \times 10^{-9} \times 0·2319 \times 19·1 \times 10^{-5} \text{ N m} = 1·33 \text{ MN m}$$

Distributing the out-of-balance end moments and multiplying by 4, the final bending moment for the full cross-section is as shown in Fig. 5.12c. The resulting primary, secondary and total thermal stresses are included in Fig. 5.12d. Note that the final magnitude of soffit thermal stress, at 4·37 MPa, is similar to that expected from maximum live load.

Analyses of bridge superstructures using this method are straightforward and yield dependable results. Typical comparisons between theory and experiment for stresses and deflections induced by thermal loading of a

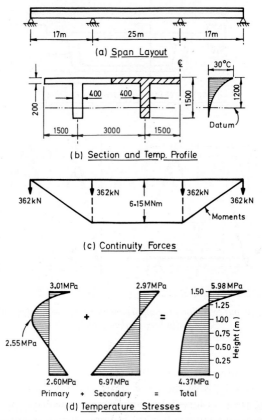

Fig. 5.12.   Thermal response of a beam-and-slab bridge.

one-quarter scale prestressed concrete box girder model are shown in Fig. 5.13. This model was subjected to simulated ambient temperature and radiation intensity variation within an environment box containing 100–375 watt infra-red light bulbs and propeller fans for cooling. Stresses and deflections were predicted from the above analysis method, using measured time histories of temperature distributions at different locations of the section.[8] Similar agreement has been obtained from *in-situ* measurements of real bridges.[9]

Non-prismatic bridge profiles: The analyses presented above assumed a prismatic bridge profile. Although solution is more tedious for non-prismatic profiles, the principles are unaltered. Consider the haunched span

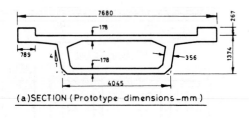

(a)SECTION (Prototype dimensions _mm)

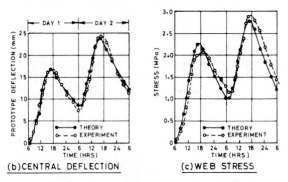

(b)CENTRAL DEFLECTION   (c)WEB STRESS

Fig. 5.13.   Theory and experiment comparisons for thermal loading of a one-quarter
scale box-girder model.

of Fig. 5.14a, assumed free to rotate at each end. The distribution of
moment of inertia $I_{(z)}$ is of the form shown in Fig. 5.14b. The curvature
distribution $\psi_{(z)}$ (Fig. 5.14c) will be approximately inversely proportional to
the section depth, and will result in the hogging displacement profile of Fig.
5.14d. To calculate $\psi_{(z)}$ it will be necessary to calculate ψ and the primary
thermal stresses at a number of sections (say, eight) along the span.

Free end rotations ϕ_1 and ϕ_2 can be found from the curvature
distribution. Assuming the span to be symmetrical about the centreline, as
in Fig. 5.14a, then

$$\phi_1 = \phi_2 = \int_0^{L/2} \psi_{(z)}\,\mathrm{d}z \tag{5.21}$$

which equals half the area under the curvature distribution, shown shaded
in Fig. 5.14c.

Equal end moments M (Fig. 5.14d, e) are required to remove the
incompatible rotations ϕ_1 and ϕ_2. Using the moment–area technique, the
M/EI diagram can be drawn in terms of the unknown M. Then, to re-

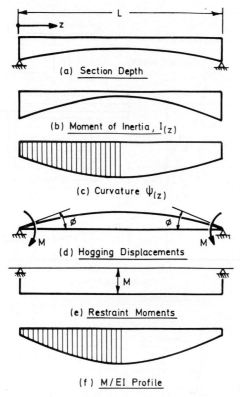

Fig. 5.14.   Thermal analysis for a haunched span.

establish compatibility, the rotation due to M must be equal and opposite to that from eqn (5.21). That is,

$$\frac{M}{E}\int_0^{L/2}\frac{1}{I_{(z)}}\,\mathrm{d}z = \int_0^{L/2}\psi_{(z)}\,\mathrm{d}z \qquad (5.22)$$

Equation (5.22) may be solved for M graphically, or directly if the distributions of $I_{(z)}$ and $\psi_{(z)}$ can be expressed mathematically. With the restraint moments for all spans calculated, the final analysis stage is to distribute out-of-balance end moments, as for a prismatic bridge.

Approximate longitudinal stress analysis: Calculation of primary thermal stresses can be tedious, particularly for complex section shapes, and design aids can be considerable benefit. Figure 5.15 summarises results for thermal

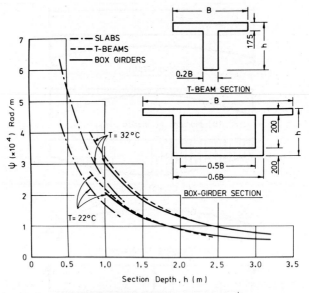

(a) UNRESTRAINED THERMAL CURVATURE, ψ

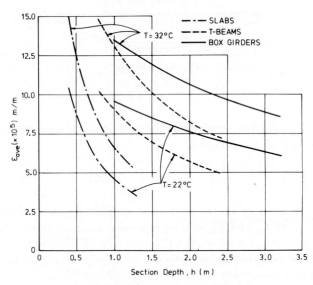

(b) UNRESTRAINED CENTROIDAL STRAIN, ε_{ave}

Fig. 5.15.   Unrestrained thermal response of different sections to design distribution of Fig. 5.5c.

curvature ψ, and thermal strain ε_{NA} at the section neutral axis, induced by the design temperature distribution of Fig. 5.5c for typical slab, T-beam and box girder sections of varying depth. The T-beam and box-girder section proportions assumed in the analyses are included in Fig. 5.15. Although the box girder represented is a single-cell box, identical results would be obtained from a multicell box whose total web width is 10% of the deck width, and whose soffit width is 60% of deck width. Two cases are considered: a deck surface temperature of $T = 32°C$ corresponding to an unsurfaced deck and $T = 22°C$ corresponding to a 50 mm blacktop thickness. Values for other blacktop thicknesses can be found by interpolation between, or extrapolation from, these two sets of values. The coefficient of thermal expansion has been assumed at $\alpha = 10^{-5}/°C$ in preparing Fig. 5.15. Results for concrete of different coefficients can be obtained by direct scaling.

It will be seen that variations in curvature with section shape are not particularly pronounced. Consequently, Fig. 15a may be used for T-beam or box-girder sections whose proportions differ from those assumed, with only small errors resulting. The variation of neutral axis strain with section type is more pronounced, mainly due to differences in neutral axis position for different sections of the same depth.

The curves of Fig. 5.15 will be sufficiently accurate for predictions of longitudinal thermal response in preliminary designs, and in many cases will suffice for final design purposes.

As an example in the use of Fig. 5.15, consider a three-span, three-lane continuous bridge with spans of 30, 40 and 30 m. The longitudinal profile is prismatic, with the section shown in Fig. 5.16a. The deck is unsurfaced, and material properties are $E_c = 30$ GPa and $\alpha = 10^{-5}/°C$.

Ignoring the fillets, and using the average thickness of the deck-slab cantilevers (200 mm), the section of Fig. 5.16a corresponds to the section proportions assumed for box girders in Fig. 5.15. Since the deck is unsurfaced, take $T = 32°C$, yielding $\psi = 1\cdot7 \times 10^{-4}$ rad/m and $\varepsilon_{NA} = 11\cdot67 \times 10^{-5}$ m/m from Figs 5.15a and 5.15b respectively. Thus, from eqn (5.17), the primary thermal stresses are

$$f_{p(y)} = 30 \times 10^9 (11\cdot67 \times 10^{-5} + 1\cdot7 \times 10^{-4} \cdot y - 10^{-5}\theta_y) \cdot \text{Pa} \quad (5.23)$$

where y is measured from the neutral axis. Equation (5.23) simplifies to

$$f_{p(y)} = 3\cdot0(1\cdot167 + 1\cdot7y - 0\cdot10\theta_{(y)}) \text{ MPa} \quad (5.24)$$

The solution for eqn (5.24) is tabulated and plotted in Fig. 5.16b. Note that, because of the different temperature profile in the deck slab above the

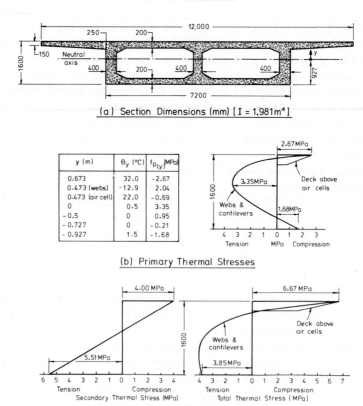

Fig. 5.16. *Thermal response of a continuous box-girder bridge.*

enclosed air cell (see Fig. 5.5c), the primary thermal stresses in the deck slab there differ from those in the webs and cantilever portions at the same level.

Since the section is prismatic, the fixed-end moments to restrain the curvature are

$$M = E_c I \psi = 30 \times 10^9 \times 1.981 \times 1.7 \times 10^{-4} \, \text{N m} = 10.1 \, \text{MN m}$$

Distribution of the fixed-end thermal moments results in a final sagging moment of 11.8 MN m in the centre span. Secondary thermal stresses from this moment are 4.00 MPa compression at the deck surface, and 5.51 MPa tension at the soffit. These are added to the primary thermal stresses of Fig. 5.16b to yield the total thermal stresses. The secondary and total thermal stresses are plotted in Fig. 5.16c.

Transverse temperature variation: In some cases, transverse variation of temperature may occur as a result of solar radiation heating one side of the bridge, or as a result of the section shape.

For these cases eqns (5.15) to (5.17) can be generalised for the case where temperature through the section varies transversely (in the x direction, Fig. 5.9) as well as vertically. In this case the temperature may be expressed as $\theta_{(x,y)}$, and the expressions become respectively

$$\psi = \frac{\alpha}{I} \int \int \theta_{(x,y)}(y - n)\, dx\, dy \qquad (5.25)$$

$$\varepsilon_0 = \frac{\alpha}{A} \int \int \theta_{(x,y)}\, dx\, dy - n\psi \qquad (5.26)$$

and

$$f_{p(x,y)} = E_c(\varepsilon_0 + \psi \cdot y - \alpha \cdot \theta_{(x,y)}) \qquad (5.27)$$

Secondary thermal stresses will depend on the value of ψ given by eqn (5.25), and may be found using the methods developed in previous sections.

Design significance of longitudinal thermal stresses: It will be seen from the examples of Figs 5.12 and 5.16 that the typical thermal stress distribution for a continuous bridge is characterised by high compressive stresses over a relatively small depth at the top of the section, and somewhat lower tensile stresses at the soffit, and also over the bulk of the section. Thus critical load combinations are likely to be those which result in high compressive stresses at the deck surface, or high tensions at the soffit. In fact, the former need not be considered, even though the combined compressive stresses due to prestress, dead and live loads and thermal loads may considerably exceed the maximum levels allowed by codes. The thermal component, which is large, is strain-induced, and ultimate compressive strains are large compared with design strain levels. There is therefore no prospect of distress through concrete crushing, at even very high thermal overloads. However, excessive soffit tensions may induce cracking, with consequent serviceability problems.

Figure 5.17 represents a three-span prestressed bridge subjected to thermal load. If the bridge is prestressed to exactly balance the dead load, then, after losses, thermal moments will add to live-load moments in the centre span, as shown in Fig. 5.17a. Cracking occurs if the total thermal plus live-load moment exceeds the cracking moment.

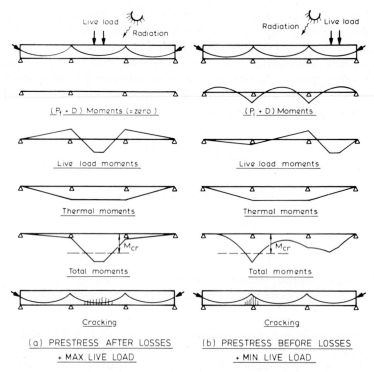

Fig. 5.17. *Potential thermally-induced cracking of a prestressed bridge (final prestress balances dead load).*

A second possible critical case is represented in Fig. 5.17b. Before all losses have occurred, prestress overbalances dead load, resulting in sagging moment at the internal supports, as shown. Sagging live-load moments are also induced at a support by live load located one span away. Combined with the thermal moment (sagging throughout) the total moment may exceed cracking at the internal support, propagating cracks from the soffit. This second case is potentially more serious than in the case represented by Fig. 17a, since the prestressing tendon will be poorly located to control any cracking that may occur. In both cases, the correct design solution will be to ensure that there is sufficient prestress to avoid cracking, though a partially prestressed approach using soffit mild steel to control cracking is feasible.

Longitudinal shears: The influence of thermal loading on shear can be divided into primary and secondary effects, as for flexure. As can be seen from Figs 5.12d and 5.16b, high primary thermal flexural stresses can develop at the level of the centroid, as well as at the soffit. Most design equations for shear required to initiate inclined web-shear cracking in prestressed members allow an increase to account for longitudinal prestress at the neutral axis (see ref. 10, for example). It is clear that primary thermal stresses will reduce the effective longitudinal precompression at the level of the neutral axis, thus tending to cause premature diagonal tension cracking.

Secondary moments from continuity effects will also induce reaction changes which may substantially increase the shear in the end spans of continuous bridges. Both of these effects must be considered in concrete bridge design.

5.2.3.2 Transverse Thermal Stress

The typical case of a box-girder section can be used to illustrate the prediction of transverse stress, 'decoupled' from longitudinal stress.

Consider the case of a transversely prestressed single-cell box-girder bridge. Under a vertical temperature distribution, the deck slab will be subjected to greater temperature variation than the soffit slab. Provided the deck slab is comparatively slender (say less than 250 mm thick) it is reasonable to assume a constant temperature gradient over the thickness.

To analyse the transverse response, the deck slab is 'removed' from the section and allowed to deform freely. As shown in Fig. 5.18, the unrestrained thermal deformation of the heated deck slab will consist of two components: a transverse average length increase of

$$\Delta = b_c \alpha (\theta_1 + \theta_2)/2 \qquad (5.28)$$

and a hogging curvature of

$$\psi = \alpha(\theta_1 - \theta_2)/h_s \qquad (5.29)$$

where θ_1 and θ_2 are the temperature increase of the top and bottom surfaces of the deck slab relative to the bottom slab, b_c is the distance between web centrelines and h_s is the deck slab thickness.

In this 'free' state, no thermal stresses are induced, since the temperature distribution is linear. However, secondary thermal stresses are induced by restraint of the deck slab elongation and hogging. The elongation can be treated as an initial-lack-of-fit problem, as shown in Fig. 5.18a: the compression P induced in the deck slabs must equal the tension T in the soffit slab and also the shear V in the webs; the sum of axial deformation of deck and soffit, plus shear and flexural deformation of webs, must equal the initial free elongation of the deck slab. Stresses induced by restraint of

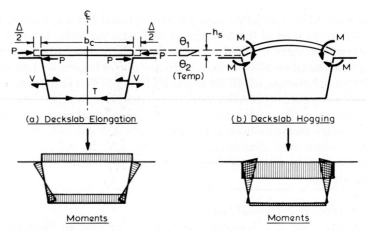

Fig. 5.18.   *Transverse moments in a box girder from restraint of deck slab thermal distortions.*

hogging (Fig. 5.18b) may be analysed by calculating the moments required to fully restrain the rotation at the web/deck joints, then distributing them round the section. Total transverse stresses are the sum of those resulting from the two components.

5.2.3.3 Biaxial Stress Fields

Theory was developed in Section 5.2.3.1 for longitudinal thermal response, on the assumption that it could be 'decoupled' from the transverse response. For slab bridges subjected to temperature variations, the biaxial strain field is significant and must be considered. Theory developed in Section 5.2.3.1 will be modified with reference to the two-way slab of Fig. 5.19.

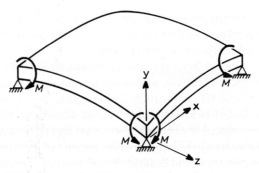

Fig. 5.19.   *Unrestrained thermal hogging of a two-way slab.*

If the slab in Fig. 5.19 is released from sufficient redundancies to make it statically determinate, it will hog freely about the x and z axes. Examination of the development of primary stress eqns (5.15) and (5.16) will show that both will independently apply to response in the x and z directions. However, since the stress-inducing strain field due to the difference between 'free' and 'final' strain states is biaxial, a modification to eqn (5.17) is required. For a biaxial strain field the constitutive relationship is

$$\left\{ \begin{matrix} f_{p(x)} \\ f_{p(z)} \end{matrix} \right\} = \frac{E_c}{1 - v^2} \begin{bmatrix} 1 & v \\ v & 1 \end{bmatrix} \left\{ \begin{matrix} \varepsilon_{(x)} \\ \varepsilon_{(z)} \end{matrix} \right\} \tag{5.30}$$

where v is Poisson's ratio for concrete. Now

$$\varepsilon_{(x)} = \varepsilon_{(z)} = (\varepsilon_0 + \psi y - \alpha \theta_{(y)})$$

as before. Substituting into eqn (5.30) yields

$$f_{p(x)} = f_{p(z)} = \frac{E_c}{1 - v} (\varepsilon_0 + \psi y - \alpha \theta_{(y)}) \tag{5.31}$$

Similarly, if the free thermal curvature ψ is restrained in both x and z directions, the restraint moments inducing secondary temperature stresses will be affected by the biaxial strain field. The constitutive relationship for moments in a biaxial field is

$$\left\{ \begin{matrix} M_{(x)} \\ M_{(z)} \end{matrix} \right\} = -\frac{E_c I}{1 - v^2} \begin{bmatrix} 1 & v \\ v & 1 \end{bmatrix} \left\{ \begin{matrix} \psi_{(x)} \\ \psi_{(z)} \end{matrix} \right\} \tag{5.32}$$

Since $\psi_{(x)} = \psi_{(z)} = \psi$, given by eqn (5.16), then

$$M_{(x)} = M_{(z)} = -\frac{E_c I \psi}{1 - v} \tag{5.33}$$

The final stage of analysis involves releasing the restraint moments at any supports where fixity is not provided, and distributing the release moments throughout the slab or shell. This can be accomplished using either a grillage or a finite element computer program. Note that, if an average value of $v = 0.18$ is assumed, eqns (5.31) and (5.33) indicate that restrained thermal stresses in a biaxial stress field will be 22% higher than for a linear beam of the same thickness subjected to the same temperature variation.

To a lesser degree, increases in thermal stresses, compared with values resulting from uniaxial considerations, will result from interaction between longitudinal and transverse stresses in the deck slab of a box girder. For transversely prestressed deck slabs, a reasonable estimate of the influence may be obtained by adding vf_t to longitudinal deck slab thermal stresses, where v = Poisson's ratio and f_t are the transverse thermal stresses, and by adding $v\Delta f_1$ to transverse deck slab thermal stresses, where Δf_1 is the bending component of the longitudinal thermal stresses. That is, $\Delta f_1 = \frac{1}{2}(f_{1t} - f_{1b})$, where f_{1t} and f_{1b} are the longitudinal thermal stresses at the top and bottom surfaces of the deck slab.

For most practical conditions, however, this interaction can be ignored since longitudinal deck thermal compressive stresses are not a critical design parameter, and the influence on peak transverse deck slab tensile stresses will be negligible. If the deck slab is conventionally reinforced in the transverse direction, and therefore subject to cracking, the concept of Poisson's ratio in the cracked zone becomes meaningless, and interaction is inappropriate.

5.2.4 Stress Analysis of Cracked Sections

The theory developed in Section 5.2.3 is based on the assumption that the material is homogeneous, isotropic and linear–elastic. As such, it can be considered to be applicable for fully prestressed sections. However, since thermal stresses in statically indeterminate structures arise mainly from bending moments which are functions of E, I and ψ, any cracking of the concrete, which will reduce the section moment of inertia, can be expected to have significant influence in reducing thermal moments. It is thus appropriate to consider a modification to the theory of Section 5.2.3 to make it applicable to conventionally reinforced bridges, and partially prestressed bridges, which will be expected to crack under design levels of loading.

5.2.4.1 Reinforced Concrete Bridges

Consider a conventionally reinforced continuous bridge which will be subject to extensive zones of cracking at normal service loads. The following assumptions are made:

(i) Concrete and steel stresses remain in the elastic range.
(ii) Uncracked concrete may support tensile stresses as high as the modulus of rupture, f_t'.
(iii) Plane sections remain plane under thermal loading.

Figure 5.20a illustrates a typical crack distribution induced by dead and live loading in a three-span bridge. The sections shown represent one-quarter of a typical beam-and-slab bridge.

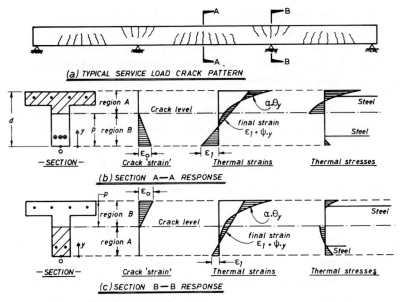

Fig. 5.20.   *Primary thermal stresses in a cracked reinforced bridge.*

Primary thermal stresses: The same approach as was adopted in Section 5.2.3.1 for uncracked sections is used. The structure is made statically determinate by the removal of internal redundancies, and the stresses resulting from non-linearity of temperature distribution are calculated. Figures 5.20b and 5.20c summarise the approach for midspan and support sections respectively. The crack penetration p is estimated from a knowledge of the current level of applied dead and live load and the past history of loading. An equivalent average crack strain ε_0 at the extreme tensile fibre is calculated:

$$\varepsilon_0 = \frac{(M_D + M_L)\bar{y}}{I_T E_c} \qquad (5.34)$$

where M_D = dead load moment, M_L = live load moment, $\bar{y}$ = distance from neutral axis to extreme tensile fibre and I_T = transformed 'all-concrete' moment of inertia of the cracked section. A linear distribution of equivalent crack strain with distance from the crack root is assumed.

As for the uncracked case, thermal stresses are induced by the difference between the free thermal strain $\alpha\theta_{(y)}$ and the final linear strain distribution $\varepsilon_1 + \psi \cdot y$.

The thermal stress distribution in the concrete and steel may now be obtained by summing the two strain diagrams. The stress distribution is discontinuous at the crack root because of cracking, as shown in Figs 5.20b and 5.20c. In the uncracked region of the section (region A) tensile stresses as high as f_t' can be sustained.

Tensile stresses cannot develop in the precracked region (region B), but compressive stresses will develop if the thermal compression strain exceeds the effective crack strain at this level. Thus, for region B, the concrete stress-inducing strain ε_c can be written

$$\varepsilon_c = \varepsilon_1 + \psi y - \alpha \theta_{(y)} - \varepsilon_{cr(y)} \geq 0 \qquad (5.35)$$

the zero value being adopted if the expression results in a tensile strain. The effective crack strain in eqn (5.35) is given by

$$\text{Section AA: } \varepsilon_{cr(y)} = \varepsilon_0 \frac{(p-y)}{p}; \ 0 \leq y \leq p \qquad (5.36a)$$

$$\text{Section BB: } \varepsilon_{cr(y)} = \varepsilon_0 \frac{(p+y-d)}{p}; \ (d-p) \leq y \leq d \qquad (5.36b)$$

Equilibrium under thermal loading requires that the axial force P and the bending moment M generated by the thermal distribution for a statically determinate system must be zero, since no external forces can be developed to resist P and M. Thus, integrating over the section:

$$P = E_c \int^A (\varepsilon_1 + \psi y - \alpha_c \theta_{(y)}) b_{(y)} \, dy$$

$$+ E_c \int^B \varepsilon_c b_{(y)} \, dy + E_s \sum_{i=1}^m (\varepsilon_1 + \psi y_i - \alpha_s \theta_i) A_{s_i} = 0 \quad (5.37)$$

$$M = E_c \int^A (\varepsilon_1 + \psi y - \alpha_c \theta_{(y)}) b_{(y)} y \, dy$$

$$+ E_c \int^B \varepsilon_c b_{(y)} y \, dy + E_s \sum_{i=1}^m (\varepsilon_1 + \psi y_i - \alpha_s \theta_i) y_i A_{s_i} = 0 \quad (5.38)$$

where m = number of layers of reinforcing steel, α_c, α_s = concrete and steel coefficients of thermal expansion respectively, E_c, E_s = concrete and steel moduli of elasticity respectively, y_i = height of ith layer of reinforcement of area A_{s_i}, with temperature change θ_i, and the integrals are over region A or B, as indicated.

The only independent variables in eqns (5.37) and (5.38) are ε_1 and ψ, and hence a unique solution exists. However, since ε_c depends on the final values for ε_1 and ψ, a direct solution is not possible, and an iterative solution technique is necessary. Note also that the maximum concrete tensile stress in region A must not exceed f_t'. Consequently, crack penetration may increase as a result of thermal loading, requiring iteration to find the final value of p.

Secondary thermal stresses: To calculate secondary thermal stresses, the restraint moments required at the ends of each span to fully restrain thermal rotations are calculated, and out-of-balance moments at joints distributed to obtain the final distribution of continuity moments.

Consider the typical span shown in Fig. 5.21a. The profiles of moment of inertia, unrestrained thermal curvature and hogging displacements for the span are given in Figs 5.21b, 5.21c and 5.21d respectively. End rotations $\phi_{\psi A}$ and $\phi_{\psi B}$ can be found by integrating the curvature using the moment–area approach, as was done in Section 5.2.3.1 for non-prismatic bridge profiles. Thus

$$\phi_{\psi A} = \frac{1}{L} \int_0^L \psi_{(z)}(L - z)\,dz \tag{5.39}$$

$$\phi_{\psi B} = \frac{1}{L} \int_0^L \psi_{(z)} z\,dz \tag{5.40}$$

To restrain these rotations, end moments M_A and M_B will be required, resulting in the moments $M_{(z)}$ of Fig. 5.21e. The distribution of restraint moments along the span must, of course, be linear.

Using the moment–area approach, the $M_{(z)}/E_c I_{(z)}$ profile is formed (Fig. 5.21f) and the end rotations induced by the fixing moments found:

$$\phi_{MA} = \frac{1}{LE_c} \int_0^L \frac{M_{(z)}}{I_{(z)}}(L - z)\,dz \tag{5.41}$$

$$\phi_{MB} = \frac{1}{LE_c} \int_0^L \frac{M_{(z)}}{I_{(z)}} z\,dz \tag{5.42}$$

For full restraint of end rotations,

$$\phi_{MA} = -\phi_{\psi A} \quad \text{and} \quad \phi_{MB} = -\phi_{\psi B}$$

Hence, it is possible to solve for M_A and M_B and, after distribution at internal joints, to find the secondary thermal moments.

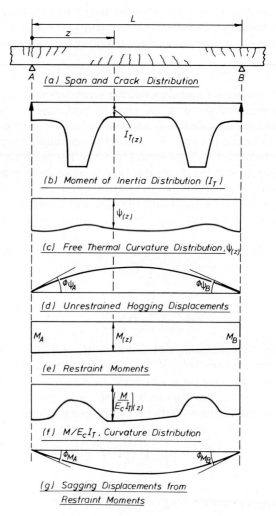

*Fig. 5.21.   Secondary thermal moments for continuous spans of a cracked reinforced
bridge.*

Secondary thermal stresses at any section z are then

$$\text{concrete: } f_{cy} = \frac{M_{(z)}(y - \bar{y})}{I_{(z)}} \tag{5.43}$$

$$\text{steel: } f_{sy} = \frac{E_s}{E_c} \frac{M_{(z)}(y - \bar{y})}{I_{(z)}} \tag{5.44}$$

where f_{cy} and f_{sy} are the concrete and steel stresses at a height y above the soffit; $\bar{y}$ is the height of the neutral axis above the soffit.

Total thermal stresses: The total thermal stresses will again be found by summing the primary and secondary stresses. However, as the secondary thermal moments will affect crack penetration and effective crack strains, they will influence the primary thermal stresses and unrestrained curvatures, and a further iteration procedure is necessary before a compatible solution is finally obtained.

Although the analytical technique developed above is clearly tedious, it is suitable for solution by computer. Thurston et al.[11] used the method to compare predicted results with those obtained from thermal loading of two-span reinforced models, where one end of the model was built in to a rigid support. Figure 5.22 presents typical comparisons between theoretical and experimental time histories of thermal continuity forces. Agreement between experiment and theory based on measured temperatures and the method developed above is good. Continuity reactions, and hence the thermal moment, are approximately 45% of the values predicted based on uncracked sections throughout the bridge, clearly indicating the significance of cracking in reducing thermal response of conventionally reinforced bridges.

Results from the experimental study indicated that primary thermal stresses for conventionally reinforced bridges were small, and that free thermal curvatures were not greatly affected by cracking. This was confirmed by analytical studies in which, using theory developed in this section, the depth of crack penetration was varied. Figure 5.23 shows the results of these studies for a typical beam-and-slab bridge, in terms of percentage change to free thermal curvature and transformed moment of inertia, from the uncracked values.

It will be seen that for crack penetrations of more than one-half the section depth (the normal cracked-section condition) theoretical thermal curvatures are very close to the uncracked value. Moment of inertia

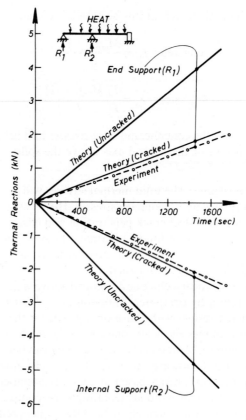

Fig. 5.22.   *Theoretical and experimental thermal reactions for a continuous cracked reinforced bridge model.*

decreased markedly with increasing crack penetration, but for normal-service load crack penetration levels, the rate of change with increasing crack penetration is small.

Simplified design approach: On the basis of these observations, a simplified design approach may be proposed for conventionally reinforced bridges.

(1) Primary thermal stresses may be ignored. Primary thermal stresses in reinforcement will normally be compressive and may conservatively be ignored when assessing maximum design tension in the reinforcement.

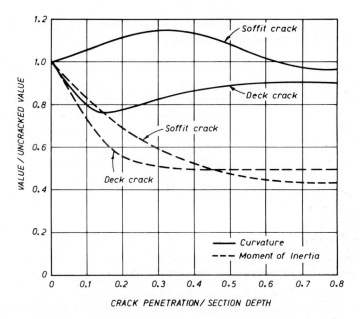

Fig. 5.23. Theoretical variation of thermal response with crack penetration for a typical reinforced slab-and-beam bridge.

Although ignoring concrete primary thermal stresses will result in underestimation of peak surface compression stresses, this is not considered important since thermal stresses are strain-induced, and hence are proportional to the tangent modulus of elasticity of the concrete. Thus, thermal compressive stresses will be reduced if any tendency to overload the concrete in compression occurs.

(2) Free thermal curvatures may be calculated on the basis of the uncracked section.

(3) Thermal continuity moments and secondary thermal stresses may be calculated from the transformed moment of inertia distribution on the basis of the expected distribution of cracking at service loads. For many applications involving prismatic bridges, sufficient accuracy may be obtained using the Branson equation,[10] averaging the values obtained for midspan and support section and assuming the average value to apply over the whole bridge length. The result will be a substantial reduction, typically more than 50% of moments compared with those based on the uncracked

section, as would be adopted for prestressed concrete bridge design using the method developed in Section 5.2.3.1.

This simplified approach will yield results to within an accuracy of about 10% of the full theoretical solution.

It should be noted that this simplified approach will be appropriate for the transverse analysis of sections, such as box girders, that may be prestressed longitudinally but conventionally reinforced transversely. For typical transverse reinforcement ratios of 1–2%, this will mean a very substantial reduction in thermally-induced transverse forces.

5.2.4.2 Partially Prestressed Bridges

Similar analytical and experimental studies to those described above have also been carried out for partially prestressed bridge sections by Thurston et al.[12] Again, a multiple iteration technique was necessary to predict the combined effects of primary and secondary thermal stresses, and again good agreement was obtained between predicted and measured response.

However, it appears that a simplified design approach, similar to that proposed in Section 5.2.4.1 for conventionally reinforced concrete sections, cannot be adopted. The simplified approach for conventionally reinforced sections depended on the observation that unrestrained thermal curvatures were almost independent of the height of crack penetration. Unfortunately a similar approximation is not valid for partially prestressed members.

Figure 5.24 shows the theoretical variation of unrestrained thermal curvature and moment of inertia with crack penetration from the soffit for a typical box-girder section (beam 1) and a typical T-beam section (beam 2). A strong dependence of curvature on crack penetration is apparent. Further, the subdivision of the bridge length into cracked and uncracked zones is not feasible until the magnitude of the thermal moment is known, and crack penetration is more dependent on the magnitude of the total applied moment than is the case for a conventionally reinforced bridge. Thus it is difficult to predict the distribution of stiffness along the bridge for use in a simplified design method.

The above arguments indicate that a simple, generally applicable method for including the influence of cracking on thermal response could be impossible to develop. On the other hand, the theory developed by Thurston et al.,[13] though predicting thermal moments with adequate accuracy, is probably too complex for use in the design of all but very large bridges.

In many design situations, partial prestressing will be adopted only to the

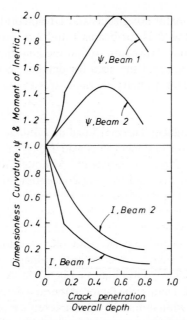

Fig. 5.24.   Theoretical variation of thermal response with crack penetration for partially prestressed sections.

extent that cracking under full live load plus thermal load would be permitted, but cracking under less severe load combinations (say dead plus live load) would be prohibited. Under these circumstances, the reduction in thermal moment resulting from cracking will generally be less than 10–15% and is unlikely to be significant when compared with uncertainties involved in material properties, such as E_c and α, which proportionately affect the thermal moment.

It is therefore recommended that bridges designed for cracking under thermal load combinations, but designed to remain uncracked under less severe load combinations, should be designed for thermal moments calculated using section stiffnesses, and thermal curvatures appropriate for uncracked sections. Thermal stresses resulting from non-linearity of the temperature profile (primary thermal stresses) can conservatively be ignored.

For bridges designed on the basis of a more liberal approach to partial prestressing, such as cracking under live load, the above recommendation is likely to be unrealistically conservative. Until more extensive design studies have been made, an indication of the change to section stiffness and unrestrained curvature for box girders and T-beam bridges can be obtained from the curves in Fig. 5.24 for beams 1 and 2 respectively.

A simplified iteration procedure to find the final thermal moments can then be carried out without the need for detailed section analysis under thermal loading. Again, primary thermal stresses may be conservatively ignored.

5.2.5 Influence of Ambient Thermal Load on Ultimate Limit State

There is a tendency for designers to consider thermal effects in bridges in terms of equivalent forces or moments. Although this is acceptable at service load levels, it can lead to a misconception of the significance of thermal loading for ultimate behaviour. For example, Fig. 5.25 illustrates the significance of thermal loading at both service and ultimate load levels. At service loads, the total effect is found by adding the thermal deformation T to the deformation induced by dead plus live load. The equivalent thermal force would then be the vertical ordinate of the shaded area between the top two horizontal lines in Fig. 5.25a. At ultimate, the same approach is adopted. That is, the factored thermal deformation ($1.7 \times T$) is added to the deformation induced by the factored service loads ($1.4 + 1.7L$). The equivalent thermal force is again the vertical ordinate of the shaded area between the top two horizontal lines in Fig. 5.25b, and is clearly of less significance than at service loads, due to the non-linearity of the force–deformation curve. For a force–deformation characteristic with a horizontal yield portion prior to failure, the ultimate thermal force may be zero, and the only significance to ultimate performance is a slight reduction in ductility or redistribution capacity. Note that factoring the equivalent

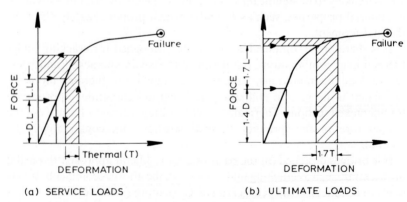

(a) SERVICE LOADS   (b) ULTIMATE LOADS

Fig. 5.25. Influence of thermal load on ultimate capacity.

thermal force, rather than deformation from the service forces in Fig. 5.25a, would erroneously predict failure under the ultimate load combination.

In all practical situations, ambient thermal loading can thus be ignored in ultimate limit states calculations without significant error.

5.2.6 Ambient Thermal Deflections

In general, thermally-induced deflections are unlikely to be significant. However, during self-supporting segmental construction, large cantilever deflections can result from thermal hogging. This can have two adverse effects. During segmental construction, it is normal practice to check deflections after each pour, and compensate in the next pour for any errors from the required vertical profile. It is clear that such corrections should not be based on deflections resulting from transient thermal response. A second effect can be the locking-in of unacceptably high deformations during the closing pour.

In Fig. 5.26a the closing pour is placed with a high thermal gradient on the bridge, inducing hogging displacements as shown. If the concrete reaches initial set with the gradient still on the bridge, cracking will occur at the deck surface at night when the gradient is removed, as shown in Fig. 5.26b. When continuity prestress is applied to the closing pour, the cracks will be closed, reintroducing the deflections existing when the section was poured. In this case local deformations at the closing pour may be high, and stresses may be locked in. However, since such stresses will be subject to creep relaxation, they will not generally be significant.

The effects of solar radiation can be substantially reduced by covering the deck with wet hessian and allowing the evaporation to cool the surface. Alternatively, the surface may be shielded by a temporary insulating or reflecting layer.

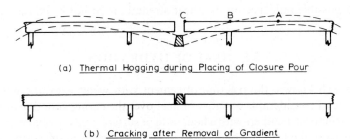

(a) Thermal Hogging during Placing of Closure Pour

(b) Cracking after Removal of Gradient

Fig. 5.26.   Thermal response during cantilever construction.

5.2.7 Designing to Avoid Thermal Stress

The examples discussed in Section 5.2.3.1 indicated that secondary thermal stresses can be of substantial magnitude, and may dominate design by increasing soffit tension stresses to unacceptable levels. In some cases, provided the durability problems associated with joints can be overcome, it may be warranted to consider adopting a structural form that eliminates these stresses. This requires a statically determinate structure so that no restraint of thermal hogging is provided.

The obvious solution, using simply supported spans, is inefficient in the way dead and live load are supported. A design which still maintains most of the desirable characteristics of continuity is the cantilever and drop-in-span construction shown in Fig. 5.27a. Although statically determinate,

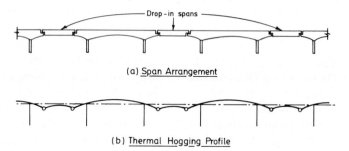

(a) Span Arrangement

(b) Thermal Hogging Profile

Fig. 5.27.   Cantilever plus drop-in span construction to minimise ambient thermal stresses.

dead and live load stresses are similar to those for an equivalent continuous structure. As shown in Fig. 5.27b, there is no restraint to thermal hogging, provided the connection between cantilever and drop-in span is hinged. In this case, thermal loading induces only primary thermal stresses and, except for a possible influence on the shear strength, need not be considered.

5.3   HEAT OF HYDRATION

As is well known, the reaction of cement with water is exothermic and liberates a considerable quantity of heat over the curing period. Although thin structural members with a high ratio of surface area to volume are able to dissipate the heat by conduction and convection from the surfaces and thus avoid significant temperature rise, the heat released by hydration in thick concrete members cannot be conducted away at a rate sufficient to avoid temperature buildup. Temperature rises at the centre of large members up to 50–60°C have been measured.[13,14]

In general, the actual temperature rise is not particularly significant in itself, as mechanical material properties are effectively independent of temperature within the range of interest. However, deformations induced by the temperature rise and subsequent cooling need to be accommodated (and hence calculated). Restraint of the free thermal expansion and subsequent contraction (for example, by placing against cured concrete, or as a result of non-linear temperature distributions through the section) can induce thermal stresses that may be of sufficient magnitude to cause cracking. Thermal cracking induced by heat of hydration can be significant for bridge diaphragms over internal piers, foundation pads and thick structural members in general.

Factors affecting peak temperature rise resulting from heat-of-hydration release are well understood. In particular, ACI Committee 209[15] has collated a large body of data and produced guidelines for predicting average thermal response of mass concrete placements.

However, to accurately predict temperature distributions, stresses and deflections, and potential for cracking, more detailed analyses are necessary, particularly when the section, while large enough to be affected by hydration effects, is still small enough to be significantly influenced by ambient fluctuations. Complete thermal analysis of concrete structures under these conditions is extremely complex. In general, heat flow is a three-dimensional phenomenon, though in many real cases it will again be sufficiently accurate to model behaviour by two- or one-dimensional heat flow.

Because of the young age of hydrating concrete, the full interactions represented by Fig. 5.1 must be considered. In particular, creep, which is insignificant for diurnal ambient thermal loading, will be a major factor in reducing thermal stresses induced by heat-of-hydration effects.

As it will be rare that detailed analyses for heat-of-hydration effects will be necessary for bridge structures, a detailed theoretical development is inappropriate here, and the interested reader is directed to ref. 16 for a more complete presentation. Where problems are expected, heat-of-hydration effects can be reduced by using chilled aggregates and ice water in the mix, and by avoiding concrete placement at high ambient temperatures. Both of these measures will reduce the rate of hydration heat release. The practice of passing cooling water through pipes embedded in the concrete is not particularly effective, as it can induce locally high tensile stresses in the vicinity of the cooling pipes. In fact, it has been shown[16] that better performance can be expected by insulating the free surfaces of large pours to reduce the rate of heat loss. Although this may increase the maximum temperature induced by heat of hydration, it reduces the maximum temperature differential between the hot centre of the concrete mass and the cooler surfaces. It is this temperature differential that is primarily responsible for inducing cracking. The decreased rate of temperature change also enhances creep relaxation of maximum stress levels.

5.4　PROPERTIES OF CONCRETE FOR THERMAL ANALYSIS

Stresses resulting from force-inducing loads such as prestress, dead and live load are comparatively insensitive to the material properties. The accuracy of stress prediction under known loads is largely dependent on the accuracy of theories for structural and section analysis. The same cannot be said for thermally-induced stresses, which result from restraint of deformation. Examination of eqns (5.1) and (5.2) shows that temperatures induced in a concrete structure by a known heat input will depend on the thermal conductivity k, the density ρ and the specific heat c. Equations (5.17)–(5.20) indicate that the magnitude of thermal stress developed from a known profile of temperature change is directly proportional to the modulus of elasticity E_c and the coefficient of thermal expansion α. In general, these material properties will not be known with any accuracy at the design stage, and, consequently, the accuracy with which predictions of thermal stress can be made is not as high as for stresses resulting from force loading. The difficulties are further compounded by the effects of creep and the temperature dependency of some of the relevant material properties.

5.4.1 Modulus of Elasticity

For normal-weight concrete, the American Concrete Institute[10] recommends the following relationship for predicting the modulus of elasticity:

$$E_c = 4730\sqrt{f_c'}\ \text{MPa} \tag{5.45}$$

where f_c' is the cylinder crushing strength in MPa. European practice, embodied in the CEB/FIP recommendations,[17] uses a different expression:

$$E_c = 9500(f_c' + 8)^{1/3}\ \text{MPa} \tag{5.46}$$

For typical concrete strengths, eqn (5.46) yields a value for E_c about 20–25% higher than that from eqn (5.45), which will be reflected in calculated values of thermal stress. It should also be noted that the value of f_c' traditionally used when calculating E_c is the specified 28-day strength. The actual strength achieved from control specimens is likely to exceed this value due to the common practice of aiming for a target strength some 20% higher than the specified strength to allow for concrete variability. It should also be borne in mind that the strength of the concrete in a structure varies with both location and time.

Evidence from testing[18,19] indicates that eqn (5.46) will produce a more realistic estimate than eqn (5.45), but is still likely to underestimate the modulus. For example, ref. 19 reports results from live load testing on a two-span box-girder bridge. Tests on site-cured concrete at the time of live load testing gave a strength of $f_c' = 34.5$ MPa. An estimate of the average E_c *in-situ* value was available from comparison of measured strains and deflections, with results predicted by a number of analyses ranging in sophistication from simple beam theory to 3-D finite-element analysis. All methods gave similar average results. The value for E_c thus calculated was compared with ACI and CEB/FIP predictions, and with experimental values from compression testing of the site-cured cylinders, giving the following results:

Tests on 200 × 100 mm cylinders $E_c = 34.6$ GPa
ACI (eqn (5.45), $f_c' = 34.5$ MPa) $E_c = 27.8$ GPa
CEB/FIP (eqn (5.46), $f_c' = 34.5$ MPa) $E_c = 33.1$ GPa
In-situ (live load tests) $E_c = 43.1$ GPa

Similar trends are reported in the literature.[18,19]

In many cases, the high *in-situ* modulus may not be significant, as the increase in tensile strength may compensate for the corresponding increase in E_c. However, where the design brief requires that residual compressions exist under all service load combinations, it is recommended that a realistic value of E_c based on the *probable* cylinder strength, using eqn (5.46), be adopted.

5.4.2 Coefficient of Linear Thermal Expansion

The coefficient of linear thermal expansion of concrete, α, can vary between $5 \times 10^{-6}/°C$ and $15 \times 10^{-6}/°C$, depending primarily on aggregate type, with andesites and limestones typically giving lowest values and quartzites typically giving highest values. Table 5.1 lists some typical average values for water-cured concrete at 20°C.

TABLE 1
Typical Coefficients of Thermal Expansion for Water-
cured Concrete made from Different Aggregate Types

Aggregate	Coefficient of thermal expansion $(10^{-6}/°C)$
Andesite	6·5
Basalt	9·5
Dolerite	8·5
Foamed slag	9
Granite	9
Greywacke	11
Limestone	6
Pumice	7
Quartzite	13
Sandstone	10

For general use, when aggregate type is not known, a value of $\alpha = 10^{-5}/°C$ can be adopted as a reasonably conservative average.

5.4.3 Thermal Properties

The designer will not in general need to be concerned with thermal properties (conductivity, specific heat, density) except in the comparatively rare cases where a heat-flow analysis is necessary to determine the design temperature gradients. In such cases, reference should be made to specialised texts (e.g. ref. 20) or, where feasible, tests for thermal properties should be carried out on the concrete that will be used on the project.

REFERENCES

1. Leonhardt, F., Kolbe, G. and Peter, J., *Beton- und Stahlbetonbau*, 7, 157–63 (July 1965).
2. Leonhardt, F. and Lippoth, W., *Beton- und Stahlbetonbau*, 10, 231–44 (Oct. 1970).
3. Priestley, M. J. N., *N.Z. Engineering*, 27(10), 213–19 (1976).
4. Prestressed Concrete Institute (Chicago) and Post-Tensioning Institute (Phoenix), *Precast segmental box girder bridge manual* (1978).
5. British Standards Institution, BS 5400, Part 2, *Steel, concrete and composite bridges—Specification for loads*, 43 pp. (1978).
6. Ministry of Works and Development, New Zealand, Highway Bridge Design Brief, CDP 701/A, 52 pp., 1978.

7. Hoffman, P. C., McClure, R. M. and West, H. H., *J. Prestressed Concrete Institute*, **28**(2), 78–97 (March/April 1983).
8. Priestley, M. J. N., *Proc. 9th Congress IABSE*, 737–46 (1972).
9. Priestley, M. J. N. and Wood, J. H., *Proc. RILEM Symposium on Testing Concrete Structures* In-situ, Vol. 1, 140–53 (Budapest, 1977).
10. American Concrete Institute, *Building code requirements for reinforced concrete*, ACI 318-83, 111 pp. (1983).
11. Thurston, S. J., Priestley, M. J. N. and Cooke, N., *Concrete International*, **6**(8), 36–43 (1984).
12. Cooke, N., Priestley, M. J. N. and Thurston, S. J., *J. Prestressed Concrete Institute*, **29**(3), 94–115 (1984).
13. Priestley, M. J. N. and Miles, J. W., *Proc. 7th Australian Road Research Board Conference*, Paper A52, 11 pp. (1974).
14. Hejnic, J., *Proc. 7th Congress FIP*, New York, Fédération Internationale de la Précontrainte, London (1975).
15. ACI Committee 209, *Designing for effects of creep, shrinkage, temperature in concrete structures*, SP-27, American Concrete Institute, Detroit, 51–93 (1971).
16. Thurston, S. J., Priestley, M. J. N. and Cooke, N., *J. American Concrete Institute*, **77**(5), 347–57 (1980).
17. Comité Euro-International du Béton/Fédération Internationale de la Précontrainte, International recommendations for the design and construction of concrete structures, Paris (1978).
18. RILEM, *Proc. International Symposium on Testing* in-situ *of Concrete Structures*, Budapest (3 vols) (1977).
19. Priestley, M. J. N., *Testing and analysis of a continuous prestressed concrete box-girder bridge*, University of Canterbury, Dept of Civil Engineering Research Report No. 77-14, Christchurch, New Zealand, 21 pp. plus Appendices (1977).
20. Neville, A. M., *Properties of Concrete*, 2nd (metric) Ed., Pitman Press, Bath (1973).

6

Membrane Enhancement in Top Slabs of Concrete Bridges

B. deV. BATCHELOR

Department of Civil Engineering, Queen's University,
Kingston, Ontario, Canada

6.1 INTRODUCTION

The design of rectangular concrete deck slabs of bridges has traditionally been based on methods developed by Westergaard[1] in which it is assumed that a wheel load is distributed over a prescribed effective width.[2,3] The slab is designed for bending effects only, the assumption being that the resulting shear capacity is adequate. The effects of any in-plane forces are thus neglected; however, it has been known for some time that these in-plane forces can have considerable effects on slab behaviour. In particular, when these forces are compressive, they have been shown to give much enhancement to slab strength.

The effects of in-plane compressive forces in the slabs are referred to as *compressive membrane action.* Such an action has been shown to affect slab response since the slab panels tend to act as arches between supporting girders, hence the term 'arching action' is sometimes used for compressive membrane action. This action is the main reason why bridge slabs designed for bending tend to fail in the punching shear mode at loads well above the predicted flexural strength.

Research conducted at Queen's University commencing in the late 1960s has established that the effect of compressive membrane action in bridge deck slabs can be predicted quite closely for certain systems. The initial theoretical and laboratory investigations have been supplemented by field studies, and the current Ontario Highway Bridge Design Code[4b] (OHBDC) prescribes the use of an empirical design method for most bridge slab systems.

The method, which recognizes the fact that considerable compressive membrane action is developed in typical bridge deck slabs, results in the use of much lower reinforcement percentages than would result from conventional design methods. Alternatively, the method could be used to assess the capacity of an existing top slab, possibly with corroded reinforcement, and, by providing a more realistic assessment of capacity, could lead to considerable savings in deck rehabilitation. It is worth noting that the theory on which the empirical method is based was used in the development of a section of the Ontario Highway Bridge Design Code[4b] dealing with the evaluation of existing deck slabs.

This chapter traces the research and testing from which the empirical design method was developed. Subsequently, the results have been confirmed in the USA and the United Kingdom. It is likely that the economies achieved and the record of good performance of slabs designed by the Ontario empirical method will lead to its adoption by many more jurisdictions in the near future.

6.2   HISTORICAL BACKGROUND

6.2.1 Shear Studies

It has been generally accepted that typical bridge deck slabs tend to exhibit punching shear failures rather than flexural failures. The punching mode of failure was first studied by Talbot[5] in his investigation of shear failures in wall and column footings. Arching action has long been regarded as a feature of beams, walls and thick slabs, but it was not until 1945 that Freysinnet[6] observed that this type of action also develops in relatively thin slabs. In a test of a trial length of a prestressed concrete runway at Orly Airport, he obtained strengths 5 to 10 times as high as would be expected from elastic analysis. Guyon[7] obtained similar results and has discussed the subject of punching shear failure in two chapters of his textbook.

Recent investigations of thin, flat slabs of buildings have confirmed the existence of compressive membrane action in these systems, particularly the effect of edge restraint. Early shear tests on slabs are numerous; these were mainly empirical studies, and statistical methods were used to establish relationships between the shear stress at failure and the parameters of the slab system.

In 1953, Hognestad[8] proposed a design equation reflecting the influence of flexural strength on the shear stress at failure of a slab. Among the other important parameters he identified were the size and shape of the loaded area, the slab span and the edge restraints. This work was extended by Elstner and Hognestad[9,10] and Whitney,[11] who all suggested refinements required to predict slab shear strength more realistically. An important common feature of these studies was the observation that slabs tend to fail in the punching mode when subjected to concentrated loads. Much care was taken to identify the 'pyramid of rupture' in the punching mode and how to use this information to improve methods for predicting punching strength. In the discussion of his work, Whitney noted that concentrations of reinforcement in a slab in the regions of supporting columns did not significantly increase slab shear strength.

Moe's investigation[12] has been regarded as an important landmark in the study of slab punching shear strength. He emphasized that it was not possible to develop a completely rational solution of the slab punching phenomenon, in view of the limited knowledge of the failure mechanism. Failing this, slab punching formulae were based on a nominal failure perimeter rather than on the actual failure mechanism.

The first rational model for punching failure was proposed by Kinnunen and Nylander[13] in 1960 with refinements by Kinnunen[14] in 1963. A mechanical model was proposed for the circular punched area of the slab on simple supports, as shown in Fig. 6.1. In this model, it is assumed that the slab portion outside the shear crack can be regarded as a rigid body rotating around a centre of rotation located at the root of the shear crack. This outer portion of the slab is assumed to be carried on an imaginary conical shell bearing on the base of the loaded area and extending to the root of the shear crack.

The criterion of failure is defined as the collapse of the conical shell. Failure is assumed to occur when the tangential strain reaches an empirical critical value. In an extension of the study, Kinnunen[14] took into account

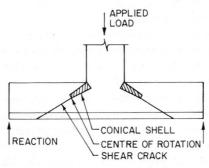

Fig. 6.1.   Kinnunen and Nylander's model for punching failure.

the dowel effects and membrane action developed in slabs provided with two-way reinforcement.

It is significant that the proposals by Kinnunen and Nylander were adopted by the Swedish State Concrete Committee in 1966[15] in their standard specifications for the design of concrete slabs supported on columns. It is also significant that after detailed examination of several existing mathematical models Hewitt[16] adopted the Kinnunen/Nylander model as the most appropriate one for extension to the study of restrained slabs, from which an important empirical slab design method was developed.[17]

6.2.2 Compressive Membrane Action in Slabs

The first documented field tests which pointed to the existence of high compressive membrane enhancement in slab strength were reported by Ockleston,[18,19] who tested a three-storey building in South Africa. The slabs exhibited strengths, the magnitude of which could not be explained by conventional theory. This started off much activity in the area of compressive membrane action in slabs. These studies were directed mainly at an understanding of the influence of this action on the flexural failure load of flat slabs, rather than its influence on punching shear strength. Significant contributions included those of Wood,[20] Brotchie,[21] Christiansen,[22] Park,[23-26] Sawczuk,[27] Jacobson[28,29] and Liebenberg.[30] Liebenberg explains clearly (Fig. 6.2) how compressive membrane action can develop in a restrained slab after cracking. Figure 6.2(a) shows that compressive membrane action can develop in an unreinforced slab. On the other hand, as shown in Fig. 6.2(c), compressive membrane action is not developed in a slab of a material having the same stress–strain relationship in compression and tension. Figure 6.2(b) shows that in a cracked slab containing reinforcement the load is resisted by flexural action and by compressive membrane action.

A very important consideration emphasized by Park[23] is that the maximum compressive membrane action develops at quite low slab deflections. Figure 6.3 shows how this action is superimposed on flexural action to give the typical load–deflection curve.

Although it is obvious how these concepts could be extended to bridge deck slabs subjected to wheel loads, it is strange that the resulting enhancement of strength was not seriously considered before the study undertaken in Ontario on such systems.

In 1965, Taylor and Hayes[31] investigated the effect of horizontal restraints on slab punching strength and reported increases of up to 60% in

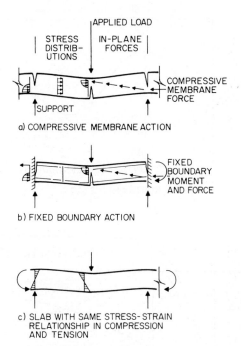

Fig. 6.2. Development of compressive membrane action.

punching strength, depending on the types of slab edge restraints. In the same year, Young[32] conducted the first study at Queen's University, Ontario, which was due to set the stage for research at that institution on membrane effects in bridge deck slabs. Subsequently, Tong and Batchelor[33-36] and Batchelor and Tissington[37,38] concluded detailed model studies of reinforced concrete bridges with square slab panels. Batchelor and Tissington[38] reported on the important non-dimensional parameters to be simulated in punching shear studies of bridge slabs. These are the flexural strength, V_{flex}, based on the yield line theory, and the slab reinforcement index, $\omega = \rho f_y / f_c'$, where ρ is the reinforcement ratio, f_y is the yield strength of the reinforcement and f_c' is the cylinder compressive strength of concrete. As shown in Fig. 6.4, it is possible with the appropriate combination of these parameters to plot all the results of slabs conducted on varying direct scale models. The transition from flexural to punching shear failure is also defined.

A great deal of work on compressive membrane enhancement in slabs was reported at the 1971 ACI Symposium.[39] This included reports by Tong

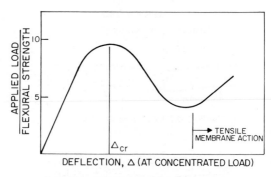

a) FULLY RESTRAINED SLAB

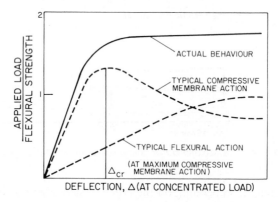

b) PARTIALLY RESTRAINED SLAB

Fig. 6.3.   Components of load–deflection curves of slabs.

and Batchelor,[34] Aoki and Seki,[40] Brotchie and Holley[41] and Hopkins and Park,[42] all of which concentrated mainly on membrane action in square slab panels. The pioneering work at Queen's University on square bridge deck slabs was sponsored by the Ontario Ministry of Transportation and Communications. This Ministry later funded the extension of the work to composite steel/concrete bridge systems from which the present OHBDC empirical design method was developed. In 1972 the Ministry acquired its first full scale test vehicle, and in 1979 Csagoly[43] reported on what was to be the beginning of an extensive field testing programme, including the testing of deck slabs. Generally, the slabs exhibited strengths well beyond their flexural capacities, and the theoretical and experimental studies at Queen's University were confirmed.

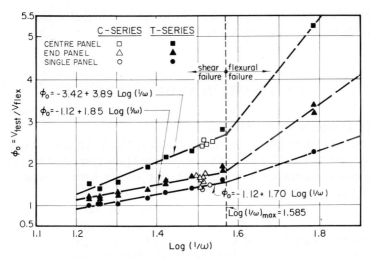

Fig. 6.4. Slab parameters for model punching studies.

6.2.3 Slabs of I-beam Bridges

Over the period 1946–53, the University of Illinois[44–46] published results of laboratory investigations of I-beam bridges. The test specimens included bridges with deck slabs which were made composite or non-composite with supporting steel beams. It was concluded that the design of composite deck slabs was much too conservative. A similar set of results was reported by Thomas and Short[47] in 1952. Since that time the major research on the behaviour of bridge deck slabs has been conducted in Ontario, and this will be discussed in some detail in the next section. It must be mentioned that the Ontario findings and applications have been of such significance that they have been tested and confirmed in other jurisdictions in the USA (New York and Texas) and in the United Kingdom (Northern Ireland).

All the investigations of bridge deck slabs have led to the general conclusion that when the edges of a slab panel are restrained against horizontal movement, the development of compressive membrane action in the system leads to enhancement in the strength of the panel. Furthermore, the punching mode of failure, rather than the flexural mode, is predominant in typical bridge deck slabs subjected to concentrated loads. The many variables inherent in a practical situation present difficulties in developing universal design approaches which take all the variables into account; therefore, recourse has been made to empirical design methods. Hewitt and Batchelor[17] first succeeded in quantifying the amount of compressive membrane enhancement to the extent that it can be used reliably in predicting slab failure strength.

6.3 THEORETICAL DEVELOPMENTS

6.3.1 Pioneer Studies in Ontario

Investigations at Queen's University, sponsored by the Ontario Ministry of Transportation and Communications, led the Ministry to conclude that the traditional methods for bridge deck slab design were so wasteful of reinforcement that they could no longer be tolerated. It was shown that even with considerable reduction of reinforcement, including its complete elimination, satisfactory safety factors can be anticipated in the predominant punching mode.

Batchelor *et al.*[17,48,49] have reported, after extensive study, that a modification of the model proposed by Kinnunen and Nylander[13] was found to be the most reliable model for predicting the punching strength of typical bridge deck slabs. Hewitt[16] showed that the basic Kinnunen and Nylander model gave good predictions of the punching strengths of simply supported slabs having their parameters within certain ranges. The model developed[17,49] for reliably predicting punching strength is shown in Fig. 6.5. It incorporated edge restraints due to membrane and bending forces, F_b and M_b respectively.

In Fig. 6.5, the outer portion of the slab, bounded by the shear crack and by two radial cracks subtending the angle β, is considered to be loaded through a compressed conical shell that develops from the perimeter of the loaded area to the root of the shear crack. The conical shell is assumed to have the section shown in Fig. 6.5(a) and a thickness varying in such a manner that the compressive stresses at the intersection with the loaded area and at the root of the shear crack are approximately equal. The criterion of failure used is the same as that previously described in connection with Kinnunen and Nylander's basic model.

The element shown in Fig. 6.5(b) is acted upon by the external force $P\beta/2\pi$ and by the following forces:

(a) the oblique compressive force $T\beta/2\pi$ in the compressed conical shell;

(b) horizontal forces in the reinforcement at right angles to the radial cracks, with resultants R_1;

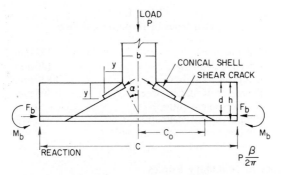

a)   SECTION SHOWING BOUNDARY FORCES

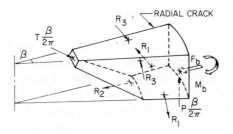

b)   FORCES ON SECTOR ELEMENT

Fig. 6.5   Model of slab at failure in punching.

(c)   horizontal forces in the reinforcement crossing the shear crack, with resultants R_2;

(d)   horizontal tangential compressive forces in the concrete, with resultants R_3;

(e)   the boundary restraining forces F_b and M_b (per unit length); F_b is assumed to act at the level of the tensile reinforcement at the boundary.

A computer program has been developed[48] using the modified Kinnunen/Nylander model to give close predictions of slab punching strengths in reported tests, provided that slab parameters fall within certain bounds defined by the equivalent span/depth ratio $C/d = 4\text{--}17$, reinforcement index $\omega = 0.05\text{--}0.45$ and a flexural factor $V_{\text{flex}}/(bd\sqrt{f_c'}) = 4\text{--}20$. The quantity C (Fig. 6.5(a)) represents the diameter or equivalent diameter of the slab; c is the perimeter of the loaded area; and d is the effective depth of the slab.

TABLE 6.1
Properties of Hypothetical Slab

Thickness, h	178 mm
Effective depth, d	140 mm
Diameter of slab, C	1 830 mm
Diameter of loaded area, b	305 mm
Reinforcement ratio, ρ	0·010
Reinforcement yield stress, f_y	310 MPa
Concrete compressive strength, f_c'	34 MPa

6.3.2 Influence of Boundary Forces

The next step was to apply the modified Kinnunen/Nylander model to a hypothetical slab in order to determine the effect of variation of boundary restraint on punching load. The characteristics of this slab are given in Table 6.1.

Values of F_b and M_b were varied over the range 0–1400 kN/m and 0–53 kN.m/m respectively, and the results obtained are shown plotted in Fig. 6.6. It is seen that the theory predicts a considerable increase in punching strength with increasing values of boundary forces.

The modified model (Fig. 6.5) was then applied to specimens used in reported tests in which the boundary conditions were known or could be deduced. This included a set of tests on unbonded prestressed concrete slabs reported by Scordelis, Lin and May,[50] for which the model gives excellent predictions of failure loads.

6.3.3 Concept of Restraint Factor

It is difficult, if not impossible, to establish precise values of the boundary restraint forces for different bridge deck slab systems, the boundary

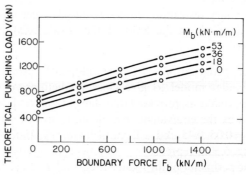

Fig. 6.6. Variation of punching load with boundary forces.

conditions and parameters of which may vary widely even for slab panels within the same structure. In order to deal with this uncertainty, Hewitt and Batchelor[17] proposed the use of a restraint factor, η, which expresses the actual values of F_b and M_b in terms of the maximum possible values $F_{b(max)}$ and $M_{b(max)}$. These maximum quantities are calculated using idealized slab displacements reported by Brotchie and Holley,[41] as shown in Fig. 6.7. The following values of $F_{b(max)}$ and $M_{b(max)}$ are obtained from Fig. 6.7, assuming that $k = 2/3$ and $f_{max} = 0.85 f_c'$:

$$F_{b(max)} = C_b - T_b \tag{6.1}$$

$$M_{b(max)} = T_b(2d - h) - C_b(d - 13h/16 - 3\delta/32) \tag{6.2}$$

in which δ represents the deflection under the load at punching failure.

Obviously, $F_{b(max)}$ and $M_{b(max)}$ would be developed only with rigid slab boundaries. For practical situations (non-rigid boundaries) it is assumed that the actual values of boundary restraining forces can be expressed as follows:

$$F_b = \eta F_{b(max)} \tag{6.3a}$$

$$M_b = \eta M_{b(max)} \tag{6.3b}$$

where $\eta \leq 1.0$. It should be noted that η is an empirical factor, the value of which can be determined for a particular system by means of testing. Also, η

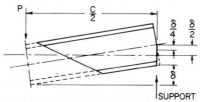

a) GEOMETRY OF DISPLACEMENT

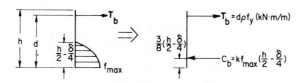

b) ASSUMED MAXIMUM BOUNDARY FORCES

Fig. 6.7.   Idealized displacements and boundary forces in fully restrained slab.[41]

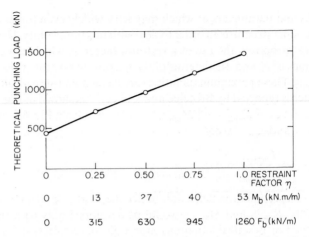

Fig. 6.8.   *Punching load vs restraint factor for hypothetical slab (Table 6.1).*

would have a value of zero for simply supported slabs, and a value of 1·0 for fully restrained slabs. A second computer program has been reported by Batchelor *et al.*[48] which enables the calculation of punching load, for different values of η, in a slab system for which punching strength has been determined from testing. This program was used to develop curves for punching load versus restraint factor, η, for different values of F_b and M_b in the hypothetical slab described in Table 6.1. The results are shown plotted in Fig. 6.8.

It can be seen from Fig. 6.8 that in the hypothetical slab the enhancement of slab punching strength can be increased by a factor of about 3, depending upon the edge conditions. Hewitt and Batchelor[17] calculated values of η in 22 tests conducted by them, and another 25 tests reported by Taylor and Hayes[31] and Aoki and Seki.[40] Table 6.2 summarizes the η-values obtained for ranges of C/b and different values of ω. It is important to note that the values of η given in Table 6.2 have been confirmed in a large number of subsequent laboratory and field tests which are discussed later.

TABLE 6.2
Calculated η-Values

C/b	ω	η
$\leq 6{\cdot}0$	0·10	0·50
6·0–9·0	0·20	0·25

Attempts have been made by Csagoly, Holowka and Dorton[51] and more recently by Kirkpatrick, Rankin and Long[52] to explain quantitatively the effect of edge restraints in deck slabs. In the latter study, the punching strength of the fully restrained slab is developed with the use of an idealized model proposed by McDowell, McKee and Sevin[53] for use in arching studies of masonry walls. The calculated maximum arching moment is then modified to reflect an effective steel reinforcement percentage, rather than an explicit restraint factor. However, unlike Hewitt and Batchelor,[17] Kirkpatrick et al. have not allowed for the effect of varying reinforcement ratio (or index)—an important consideration in the design and evaluation of deck slabs of bridges.

Extensive studies funded by the Texas Transportation Department have been recently completed at the University of Texas at Austin.[54-56] Although the reports were not officially available at the time of writing, the author was informed by private communication from one of the investigators that the studies indicate that the values of η recommended by Hewitt and Batchelor[17] are on the conservative side.

6.3.4 Non-linear Finite Element Studies
Cope et al.[57,58] have reported the results of analytical studies of restrained slab strips and of top slab panels using non-linear finite element methods. Their approaches follow the development of concrete cracking and steel yielding with incrementally increasing loads. Their analytical predictions reflect those described above, and give additional information on structural response under working loads.

Failure loads predicted by Cope et al. compared very closely with those reported in tests of slab panels.[17,38,52] This is surprising, since the non-linear method used does not model punching failure explicitly. The authors state that failures in the analytical models were due to crushing of the concrete beneath the load. They speculate that this could have triggered punching failures in the physical tests. Non-linear finite element models may provide useful information on details of structural mechanisms, but are too costly for direct use in design.

6.4 LABORATORY AND FIELD STUDIES

6.4.1 Laboratory Studies
The initial laboratory studies conducted at Queen's University on direct scale models with square slab panels[32,33-36,37,38] confirmed that punching

strength can be satisfactorily studied with direct models of slabs down to $\frac{1}{16}$ scale. These tests also confirmed the existence of high levels of compressive membrane action in typical square slabs, and provided useful information on the effects of edge members. However, subsequent work concentrated on the rectangular deck slabs of composite beam–slab systems, these being the most widely used types of bridge in Ontario.

The first major study of the punching strength of the slabs of composite steel/concrete bridges is described in detail in ref. 16. The analytical aspects of this study have already been discussed. The experimental phase of the study was done on $\frac{1}{8}$ scale models of prototype three-beam and four-beam bridges spanning 24·4 metres. Some deck slab panels were reinforced with conventional orthotropic reinforcement, while others were reinforced with isotropic reinforcement varying from zero up to a maximum of 0·6%. Shear connector behaviour and dead load stresses appropriate to unshored construction were simulated. Other variables included dead load stresses, reinforcement location and concrete strength. The detailed modelling considerations of the study presented some interesting aspects of the use of models for ultimate conditions. The main model and prototype dimensions are shown in Fig. 6.9.

The slab panels were tested statically to failure under concentrated loads applied between adjacent beams through areas representing the contact area of the pneumatic tyres of large trucks. Table 6.3 indicates that a total of nine bridge specimens were tested statically, i.e. a total of 108 slab panels. All the reinforced slabs failed mainly in the punching mode of the type shown in Fig. 6.10.

Typical load–deflection plots are shown in Fig. 6.11, which indicates the increased ductility but generally lower punching strengths of the isotropically reinforced slabs compared to the relatively heavily reinforced orthotropically reinforced slabs.

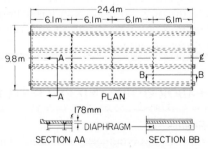

Fig. 6.9. Description of test specimens.

TABLE 6.3
Characteristics of Slab Test Panels

Bridge number	Reinforcement				Special tests and remarks
	Strip 1	Strip 2	Strip 3	Strip 4	
1	Ortho[a]	Ortho	Ortho	Ortho	
2	Ortho	Ortho	Ortho	Ortho	Full dead load compensation (all panels)
3	Ortho	Ortho	Ortho	Ortho	Dead load compensation in some panels
4	Ortho	Ortho	Ortho	Ortho	Full dead load compensation (all panels)
5	Ortho	Zero	0·4%	0·2%	
6	0·6%[b]	0·2%	0·6%	M-D	Strips 2 and 3 subjected to hogging moment
7	0·4%	0·2%	Zero	0·6%	
8	0·4%	Zero	M-D[c]	0·2%	
9	0·4%	0·2%	Zero	0·6%	

Notes
[a] Ortho indicates orthotropic reinforcement.
[b] % indicates isotropic reinforcement in both faces.
[c] M-D indicates isotropic reinforcement at mid-depth only.

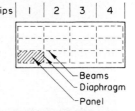

Strip designation

An interesting plot of average punching loads versus reinforcement percentage of similar isotropically reinforced slab specimens is shown in Fig. 6.12. Also indicated on the vertical axis is a scale of multiples of the design wheel load. This plot clearly indicates the high factors of safety against failure of all the slab panels, including those in which no reinforcement was provided.

Hogging moments were introduced into the deck slabs by moving the bridge supports towards midspan and loading the resulting cantilever portions until cracking occurred in the slab in the hogging moment region. The resulting slab punching loads were not markedly reduced from those obtained from tests in positive moment regions.

Batchelor *et al.* extended their investigation to repeated load studies[59] of orthotropically reinforced as well as isotropically reinforced decks (0–0·6%). The models used in these tests were similar to those used in the

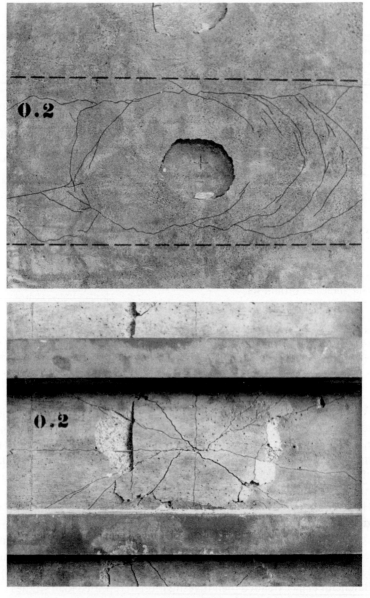

Fig. 6.10.   Typical punching failures. (a) Panel with 0·2% isotropic reinforcement.

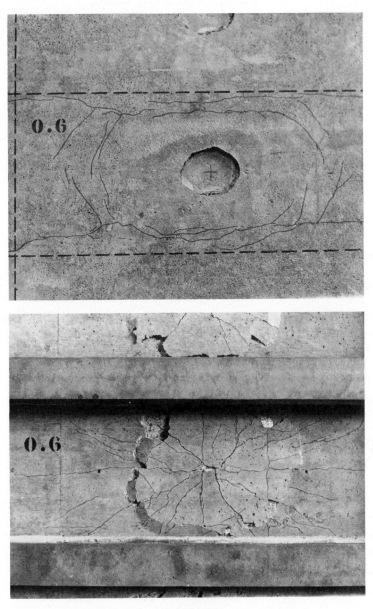

Fig. 6.10—contd. (b) Panel with 0·6% isotropic reinforcement.

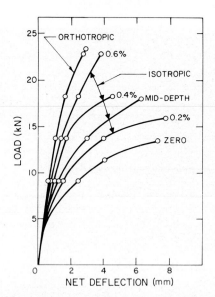

Fig. 6.11.   *Typical load–deflection curves for slabs of four-beam bridge model.*

static studies. It was concluded that the life of the slabs with isotropic reinforcement as low as 0·2% is satisfactory for the load ranges to which bridge deck slabs in Ontario are subjected.

In 1979, Csagoly[43] reported tests on circular reinforced concrete slabs surrounded by steel rings of various cross-sections which were designed to provide restraints corresponding to values of restraint factor, η, varying from 0·50 to 0·75. The specimens failed in punching under concentrated

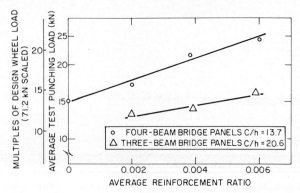

Fig. 6.12.   *Variation of punching load with reinforcement ratio for isotropic slabs.*

loads which were close to predictions using the computer program reported in ref. 48. Furthermore, the predictions were on the conservative side.

The publication, in 1979, of the first edition of the Ontario Highway Bridge Design Code[4a] contained a special section providing prescriptions for a simple empirical design of typical bridge slabs. A minimum of 0·3% isotropic reinforcement in each face was prescribed for a wide range of bridge slabs conforming to certain conditions. This evoked much interest internationally, and led to a number of laboratory studies similar to those conducted by Batchelor and Hewitt. For example, Beal,[60] using scale model studies, confirmed the existence of considerable reserve strength in concrete deck slabs due to arching action, even in those slabs containing no reinforcement.

Kirkpatrick et al.[52] reported results from tests of scale model specimens similar to the type previously investigated in Ontario and New York. This study also confirmed the findings of the earlier reports. However, Kirkpatrick et al. gave the mistaken impression that the Ontario Highway Bridge Design Code does not provide for reinforcement percentage other than the 0·3% specified for new construction. As already pointed out, an entire section of that code deals with the evaluation of existing structures. This section provides useful charts for assessing slab punching strength on the basis of girder spacing, slab thickness, concrete compression strength and average tensile reinforcement ratio at midspan of a panel. A curve developed from the computer program[48] for punching strength versus span/thickness is shown in Fig. 6.13.

The tests recently completed in Texas, USA,[54–56] have provided general confirmation of the Ontario empirical slab design method. These tests included full-scale laboratory specimens of a 50 ft (15·2 m) span bridge under static and repeated loading.

6.4.2 Field Studies

Following the very favourable results obtained in deck slab studies at Queen's University, and the very strong evidence provided for the reduction of deck slab reinforcement, the Ontario Ministry of Transportation and Communications decided to build a full-scale experimental bridge in Conestoga, Ontario, in 1975. This structure[61] incorporated other research findings besides those on deck slabs. Some panels of the deck slab were provided with conventional orthotropic reinforcement, and others contained isotropic reinforcement varying from 0·1% to 0·6%. The comprehensive testing of the bridge confirmed the validity of the empirical design method and provided justification for the inclusion of the method in the first edition[4a] of the Ontario Bridge Design Code. Tests on the Conestoga bridge in 1984[62] have shown that the deck slab still continues to perform satisfactorily.

The Ontario Ministry of Transportation and Communications has conducted extensive field testing of deck slabs in service. The test procedures used, and the results, have been reported by Bakht and Csagoly.[63] The reported tests[51] covered a total of 28 bridges. Nine bridges contained non-composite decks, another nine had decks composite with steel girders; eight bridges were of reinforced concrete T-beam construction, and the remaining two were decks constructed compositely with

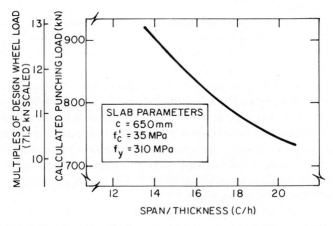

Fig. 6.13.   Punching load vs span/depth ratio ($\rho = 0.002$, $\eta = 0.50$).

prestressed concrete girders. The average values of the restraint factor, η, were 0.43 for the non-composite systems, 0.93 for the composite steel/concrete decks, 0.78 for the reinforced concrete T-beam system and 0.83 for the decks on prestressed concrete beams. In the latter system it was shown that the required compressive membrane action can be developed without the use of intermediate diaphragms in the structure. This extensive field study confirmed conclusively that a value of $\eta = 0.5$ is conservative for composite construction. Ongoing testing of bridge decks by the Ontario Ministry of Transportation and Communications for evaluation purposes has generally confirmed strong arching effects in typical bridge deck slabs. Field tests are also in progress in the USA with essentially similar findings. These studies will probably appear in the literature in the near future.

6.5 DESIGN CODE PROVISIONS

The first edition of the Ontario Highway Bridge Design Code[4a] specified an *acceptable* empirical deck slab design method based on the extensive theoretical and experimental investigation described above. A minimum of 0·3% isotropic reinforcement was prescribed in each face of the slab; that is, four layers of reinforcement, with each layer being at least 0·3% of the effective cross-sectional area of the concrete, as shown in Fig. 6.14(b). This is 50% greater than the amount of reinforcement (0·2%) recommended by Batchelor and Hewitt[17,48] in their previous studies. The Ontario Code[4a] allowed the use of the empirical method only if all the following conditions were met:

(a) Slab span (girder spacing) does not exceed 3·7 m.

(b) The slab extends at least 1·0 m beyond the exterior girder, or a kerb of equivalent cross-section is provided.

(c) The spacing of reinforcement in each layer does not exceed 300 mm.

(d) Intermediate diaphragms or cross-frames are provided at a maximum spacing of 8·0 m.

(e) Diaphragms or cross-frames are provided at all supports.

(f) The skew angle of the slab system does not exceed 20°.

(g) The ratio of girder spacing to slab thickness does not exceed 15, with the minimum slab thickness being 190 mm.

The second edition[4b] of the Ontario Highway Bridge Design Code, published in 1983, makes the empirical design method *mandatory*, provided certain conditions—less stringent than those in the first edition—are met. In particular, the conditions regarding skew are relaxed, as shown in Fig. 6.14(a), with a doubling of reinforcement in the support regions. The empirical method is now allowed for slabs supported on concrete girders without intermediate diaphragms. The minimum slab thickness has been increased to 225 mm purely from considerations of durability. The increased thickness guarantees even higher levels of compressive membrane enhancement of slab strength.

Charts are provided in both editions of the Code[4a,4b] for evaluating deck slabs not conforming to specified criteria. These charts give failure loads for composite and non-composite slabs with various parameters, and were derived from the theoretical model and computer programs described by Batchelor *et al.*[48]

Since the appearance of the Ontario Bridge Code in 1979 a large number of bridges have been constructed in which the empirical slab design method

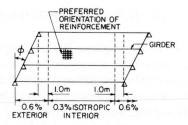

a) SUMMARY OF REINFORCEMENT
   SPECIFICATIONS (46)

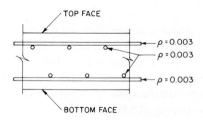

b) REINFORCEMENT IN INTERIOR
   PANEL

Fig. 6.14.   Deck slab reinforcement in OHBD Code.[4b]

was used. The savings resulting from the empirical design method are considerable. Bakht and Markovic[62] have reported savings of 35–40% in slab reinforcement, depending on girder spacing (up to 3·0 m). They have also shown how the Ontario method can be modified to effect even greater savings in areas where protection of reinforcement against deicing salts is not necessary.

6.6   CONCLUSIONS

The exhaustive theoretical and experimental studies described in this chapter have led to the conclusion that it is wasteful to design bridge deck slabs by conventional elastic methods.[1,2] The empirical deck slab design method now in wide use in Ontario[4b] is simple to apply and results in considerable saving in reinforcement, without compromising safety. The method recognizes the fact that due to the high level of compressive membrane action bridge slabs will fail in the punching rather than in the flexural mode. The research in Ontario has been duplicated elsewhere, with similar findings. It is understood that the empirical method is being adopted in Texas in the near future, and it is likely that other American states will follow suit.

In view of the economy of the empirical design method, the satisfactory performance of deck slabs in service designed by this method, and the confirmation of the method by independent investigators, it is likely that the method will soon have wide acceptance in North America and elsewhere.

REFERENCES

1. Westergaard, H. M., *Public Roads*, **11**(1), 1–23 (1930).
2. American Association of State Highway Officials, *Standard Specifications for Highway Bridges*, 13th edn, Washington, DC (1983).
3. Canadian Standards Association, *Design of Highway Bridges*, CSA Standard S6-M78, Rexdale, Ontario, Canada (1978).
4a. Ontario Ministry of Transportation and Communications, *Ontario Highway Bridge Design Code*, 1st edn, Downsview, Ontario (1979).
4b. Ontario Ministry of Transportation and Communications, *Ontario Highway Bridge Design Code*, 2nd edn, Downsview, Ontario (1983).
5. Talbot, A. N., *Reinforced concrete wall footings and column footings*, University of Illinois, Eng. Exp. Stn, Bulletin 67 (1913).
6. Harris, A. J., *Proc. Inst. Civ. Engs*, **6**, 45 (1957).
7. Guyon, Y., *Prestressed Concrete*, Vol. II, John Wiley and Sons, New York (1960).
8. Hognestad, E., *Proc. Amer. Conc. Inst.*, **50**, 189 (1954).
9. Elstner, R. C. and Hognestad, E., *An investigation of reinforced concrete slabs failing in shear*, Univ. of Illinois, Dept of Theoretical and Applied Mechanics (1953).
10. Elstner, R. C. and Hognestad, E., *Proc. Amer. Conc. Inst.*, **53**, 59 (1956–57).
11. Whitney, C. S., *Proc. Amer. Conc. Inst.*, **54**, 265 (1957).
12. Moe, J., *Shearing strength of reinforced concrete slabs and footings under concentrated loads*, Portland Cement Assn, R&D Labs, Bull. D.47 (1961).
13. Kinnunen, S. and Nylander, H., *Trans. Royal Inst. Technology*, Stockholm, No. 158 (1960).
14. Kinnunen, S., *Trans. Royal Inst. Technology*, Stockholm, No. 198 (1963).
15. Swedish State Concrete Committee, Draft specification for the design of concrete slabs supported on columns, with extracts from comments, CEB, *Bull. d'Information*, **57** (1966).
16. Hewitt, B. E., *An investigation of the punching strength of restrained slabs with particular reference to the deck slabs of composite I-beam bridges*, PhD Thesis, Queen's University at Kingston, Ontario, Canada (1972).
17. Hewitt, B. E. and Batchelor, B. deV., *J. Str. Div., Amer. Soc. Civ. Engs*, **101**(ST9) (1975).

18. Ockleston, A. J., *Struct. Engineer*, **33**, 304 (1955).
19. Ockleston, A. J., *Struct. Engineer*, **36**, 197 (1958).
20. Wood, R. H., *Plastic and Elastic Design of Slabs and Plates*, Ronald, New York (1961).
21. Brotchie, J. F., *J. Inst. Engs Australia*, **35**, 292 (1963).
22. Christiansen, K. P., *Struct. Engineer*, **41**, 261 (1963).
23. Park, R., *Proc. Inst. Civ. Engs*, **28**, 25 (1964).
24. Park, R., *Mag. Conc. Res.*, **16**, 139 (1964).
25. Park, R., *Mag. Conc. Res.*, **17**, 29 (1965).
26. Park, R., Membrane action at the ultimate load of laterally restrained concrete slabs, CEB, *Bull. d'Information*, **58**, 135 (1966).
27. Sawczuk, A., Flexural mechanics of reinforced concrete, *Proc. Int. Symp. Amer. Soc. Civ. Engs and Amer. Conc. Inst.*, Miami, 347 (1964).
28. Jacobson, A., *Membrane flexural failure modes of square horizontally restrained concrete slabs*, DSc Thesis, Mass. Inst. Tech. (1965).
29. Jacobson, A., *J. Str. Div.*, *Amer. Soc. Civ. Engs*, **93**(ST5), 85 (1967).
30. Liebenberg, A. C., *Arch action in concrete slabs*, South African Council for Sci. and Ind. Res., R234, Nat. Bldg Res. Inst., Bull. 40 (1966).
31. Taylor, R. and Hayes, B., *Mag. Conc. Res.*, **17**, 39 (1965).
32. Young, D. M., *The strength of two-way slabs with fixed edges subjected to concentrated loads*, MSc Thesis, Queen's University at Kingston, Ontario, Canada (1965).
33. Tong, P. Y., *An investigation of the ultimate shear strength of two-way continuous slabs subjected to concentrated loads*, PhD Thesis, Queen's University at Kingston, Ontario, Canada (1969).
34. Tong, P. Y. and Batchelor, B. deV., American Concrete Inst., *Special Pub. SP-30*, 271 (1971).
35. Batchelor, B. deV. and Tong, P. Y., Ontario Ministry of Transportation and Communications, *Res. Rep. No. RR167* (1970).
36. Batchelor, B. deV., On direct model studies of the punching strength of two-way bridge slabs. Experimental analysis of instability problems on reduced and full scale models. *RILEM Int. Symp.*, I, Buenos Aires (1971).
37. Tissington, I. R., *A study of partially restrained two-way bridge slabs subjected to concentrated loads*, MSc Thesis, Queen's University at Kingston, Ontario, Canada (1970).
38. Batchelor, B. deV. and Tissington, I. R., *J. Str. Div.*, *Amer. Soc. Civ. Engs*, **102**(ST12), 2315 (1976).
39. American Concrete Institute, *Proc. Symp. on Cracking, Deflection and Ultimate Load on Concrete Slab Systems*, SP-30 (1971).
40. Aoki, Y. and Seki, H., *ACI Special Pub. SP-30*, 103 (1971).
41. Brotchie, J. F. and Holley, M. J., *ACI Special Pub. SP-30*, 345 (1971).
42. Hopkins, D. C. and Park, R., *ACI Special Pub. SP-30*, 223 (1971).
43. Csagoly, P., Design of thin concrete deck slabs by the Ontario bridge design code, Ontario Ministry of Transportation and Communications, *Report SRR-79-11* (1979).
44. Newmark, N. M., Seiss, C. P. and Penman, R. R., *Studies of slab and beam highway bridges, Pt I*, University of Illinois, Eng. Exp. Stn, Bull. 363 (1946).

45. Newmark, N. M., Seiss, C. P. and Peckham, W. M., *Studies of slab and beam highway bridges, Pt II*, University of Illinois, Eng. Exp. Stn, Bull. 375 (1948).
46. Seiss, C. P. and Viest, I. M., *Studies of slab and beam highway bridges, Pt IV*, University of Illinois, Eng. Exp., Stn, Bull. 416 (1953).
47. Thomas, F. G. and Short, A., *Proc. Inst. Civ. Engs*, 1, Pt I, 125 (1952).
48. Batchelor, B. deV., Hewitt, B. E. and Holowka, M., *Load carrying capacity of concrete deck slabs*, Ontario Ministry of Transportation and Communications, *Report SRR-85-03* (1985).
49. Batchelor, B. deV., Hewitt, B. E. and Csagoly, P., *TRR Record No. 664, Bridge Engineering*, Vol. 1, 162, Washington, DC (1978).
50. Scordelis, A. C., Lin, T. Y. and May, H. R., *J. Amer. Conc. Inst.*, 55, 485 (1958).
51. Csagoly, P., Holowka, M. and Dorton, R. A., *TRR Record No. 664, Bridge Engineering*, Vol. 1, 171, Washington, DC (1978).
52. Kirkpatrick, J., Rankin, G. I. and Long, A. E., *The Structural Engineer*, 62B(3), 60 (1984).
53. McDowell, E. L., McKee, K. E. and Sevin, E., *J. Str. Div., Amer. Soc. Civ. Engs*, 915 (1956).
54. Fang, I. K., Worley, J., Burns, N. H. and Klingner, R. L., *Behavior of Ontario-type bridge decks on steel girders, Res. Rep. CTR 350-1*, Center for Transportation Research, University of Texas at Austin (1985).
55. Elling, C. W., Burns, N. H. and Klingner, R. L., *Distribution of girder loads in a composite highway bridge, Res. Rep. CTR 350-2*, Center for Transportation Research, University of Texas at Austin (1985).
56. Tsui, C., Burns, N. H. and Klingner, R. L., *Behavior of Ontario-type bridge decks on steel girders: negative moment region and load capacity, Res. Rep. CTR 350-3* (1985).
57. Cope, R. J., Rao, P. V. and Clark, L. A., *Non-linear Design of Concrete Structures*, University of Waterloo Press, Canada, 379 (1980).
58. Cope, R. J. and Edwards, K. R., *Proc. Int. Conf. Finite Elements in Computational Mechanics*, Bombay, India, Pergamon Press Ltd, 435 (1985).
59. Batchelor, B. deV., Hewitt, B. E. and Csagoly, P., *TRR Record No. 664, Bridge Engineering*, Vol. 2, 153, Washington, DC (1978).
60. Beal, D., *J. Str. Div., Amer. Soc. Civ. Engs*, 108(ST4), 814 (1982).
61. Dorton, R. A., Holowka, M. and King, J. P. C., *Can. J. Civ. Eng.*, 4(1), 18 (1977).
62. Bakht, B. and Markovic, S., Reinforcement savings in deck slab by a new design method, *Proc. Int. Colloq. on Concrete in Developing Countries*, Vol. 2, 594, Lahore, Pakistan (1985).
63. Bakht, B. and Csagoly, R. P., *Bridge testing*, Ontario Ministry of Transportation and Communications, Res. Rep. SRR-79-10 (1979).

7

Modified Concretes for Use in Bridge Structures

J. L. CLARKE

Cement and Concrete Association, Slough, UK

7.1 INTRODUCTION

Although the strength of concrete has gradually increased over the years, its basic constituents have not altered. Generally the designer of reinforced concrete will be dealing with a material made from natural aggregates with a density of about 2400 kg/m^3 and a compressive strength in the region of 25 to 40 N/mm^2. For prestressed concrete the strength is likely to be between 50 and 60 N/mm^2. In both cases the tensile strength of the concrete will be so low as to be generally ignored, tensile loads being taken by steel reinforcing bars in the first case and stressed wires, bars or cables in the second. At all times the structure relies on the reinforcement, which must be protected from corrosion by the concrete surrounding it. In the context of this chapter, 'modified' concretes should be seen as those that fall into one or more of the following categories:

(1) improved strength per unit weight;
(2) improved tensile capacity;
(3) improved protection to the reinforcement;
(4) improved resistance to chemical, environmental or mechanical attack.

In addition there are approaches which aid the construction process and which may improve the performance, either by increasing the load capacity or by enhancing the durability.

Perhaps one of the major obstacles to the adoption of modified concretes is that conventional concrete is a relatively inexpensive material. It has thus been perfectly acceptable to design what might be described as inefficient structures out of normal concrete, ones that behave adequately in terms of strength or durability but are unnecessarily conservative in some areas. For example, the provision of the required cement content to ensure adequate durability is likely to result in a concrete strength that is higher than is required from the point of view of safety. High cover to the reinforcement will increase the self-weight of members, which could be significant for long-span structures. Careful choice of a modified concrete having the specific characteristics required, possibly only being used for critical parts of the structure, could be a better solution.

Where appropriate, mention is made of the relative costs of modified and conventional concretes. These should only be taken as indicative and reflect the small usage of the materials to date: widespread adoption would tend to reduce the cost differential. In addition, costs should be considered in the context of the whole structure, taking into account construction and any maintenance or replacement that may be likely during the life of the bridge.

Where concrete compressive strengths are quoted they are cube strengths; conversions from quoted cylinder strengths have been obtained by adding $5 \, N/mm^2$ to give the equivalent cube strengths.

7.2 LIGHTWEIGHT AGGREGATE CONCRETE

7.2.1 Introduction

In bridge superstructures a major proportion of the loading is due to the self-weight. Hence there would appear to be immediate advantages to be gained from the use of lightweight aggregate concrete.

Most of the lightweight aggregates suitable for concrete are factory made and include sintered pulverised fuel ash, pelletised slag and expanded clay, shale and slate.[1] In addition, natural volcanic rocks such as pumice can be used. The resulting concretes have a range of densities and strengths. Those suitable for structural use, i.e. above about $25 \, N/mm^2$ at 28 days, will have densities between about $1400 \, kg/m^3$ and $2000 \, kg/m^3$. It is possible to achieve strengths in excess of $50 \, N/mm^2$, but this is usually done by using lightweight coarse aggregate combined with a natural sand. This obviously limits the weight saving but, typically, a sintered pulverised fuel ash aggregate with a natural sand will give a density in the region of $1900 \, kg/m^3$, a reduction of over 20% when compared with normal concrete.

The reduction in self-weight leads, in general terms, to a reduction in the size of beam required to span a given gap. Alternatively, the live load capacity for a given member size can be increased. The Concrete Society study of the economics of using lightweight aggregate concrete for bridges[2] considered structures with a span of about 25 m suitable for a dual two-lane by-pass. Lightweight concrete was used for both the precast beams and the *in-situ* deck but not for the columns, abutments, etc. On the basis of a number of quotations, the unit cost of the lightweight beams delivered to site was taken to be about 12% higher than for dense beams and the lightweight ready-mixed concrete to be about 24% dearer. Because of weight saving, the lightweight bridge required only 5 beams while the dense one required 6. Not only were fewer bearings required but they were of a slightly lower capacity. However, because of the increased transverse bending moments due to the wider beam spacing the lightweight deck was thicker.

Taken on balance, the lightweight solution showed a small cost advantage over the dense one. The report suggests that further cost savings could be made in the following areas:

(1)   reduced foundation sizes, which could be particularly important in areas with poor bearing capacity;
(2)   reduced column sizes, requiring less formwork;
(3)   lighter formwork for all in-situ concrete, and lighter falsework, leading to reduced times for erecting and dismantling;
(4)   lower erection loads, possibly leading to reduced crane capacity.

It was felt that it was impossible to place an accurate cost saving on these items and hence their financial advantages were ignored. The report concludes that savings of about 4% are likely for a 60 m span and 8% for a 200 m span. Similar savings have been noted in another study by the Concrete Society which looked at the use of lightweight concrete in multi-storey buildings.[3]

7.2.2 Applications

Although lightweight aggregate has been used for many years for bridges in North America and in Europe (as reviewed in ref. 4), there has been resistance to its use for bridges in the UK. However, there are a few examples such as the Friarton Bridge over the River Tay, which has a lightweight deck slab[5] acting compositely with twin steel box girders over nine spans with a total length of more than 800 m. The bridge was opened in 1978. More recently a small cable-stayed bridge was built at the Ealing Broadway Centre in London[6] to give access to a car park (Fig. 7.1). Since it is used only by cars the live load is low and hence the self-weight of the structure is particularly important. Sintered pulverised fuel ash was used for the coarse aggregate, with natural sand fines, which gave an average cube strength in excess of 40 N/mm². The paper reports that the additional cost of the lightweight aggregate concrete was offset by savings produced by the reduced cable sizes, mast and anchoring forces and by the lower quantity of reinforcement required. The material has been used more extensively, for example, in France where the Ottmarsheim Bridge[7] has a

Fig. 7.1. Cable-stayed lightweight aggregate bridge at the Ealing Broadway Centre, London.

170 m long main span of which the central 100 m is constructed of lightweight aggregate concrete having a cube strength of 40 N/mm². For the remainder of the span and for the piers a higher grade normal weight concrete was used, which is an efficient use of the two materials.

The Concrete Society Review[4] pays particular attention to the reduced erection loads for precast segmental bridges where the use of lightweight aggregate concrete can lead to a significant reduction in the number of segments. In one example quoted the number of segments for a precast concrete box girder bridge was reduced from 820 to 600, with significant savings being possible on the time and cost of construction. Including an additional saving on the prestressing because of the reduced self-weight, the overall saving when compared with a conventional dense concrete solution was 15%. A further area of interest is the upgradimg of existing bridges, where the substitution of a lightweight deck onto existing beams will lead to an improved load capacity, possibly combined with an increased carriageway width. For example, the deteriorating normal weight deck of the Woodrow Wilson Memorial Bridge in America was recently replaced with precast lightweight slabs, which enabled the carriageway width to be increased from about 11·6 m to about 13·4 m.[8]

At present, within the UK, the cost of lightweight aggregate concrete as delivered to site is, perhaps, 20% more than that of normal weight concrete of an equivalent grade. However, when all the factors considered above are taken into account, there is likely to be an overall cost saving. In the future, as the sources of good quality natural aggregate near to the centres of demand become worked out, leading to increased haulage costs, the use of lightweight concrete will become more attractive, particularly if there are pressures for the use of waste materials such as pulverised fuel ash and blast-furnace slag. For example, an all lightweight concrete bridge, i.e. both substructure and superstructure, is currently being built in Holland where pressure from conservationists means that the price of the aggregate is equal to that of a natural aggregate.

7.2.3 Design Considerations for Reinforced Concrete

Although lightweight concrete can readily be produced in strengths suitable for structural reinforced concrete, its mechanical properties differ somewhat from those for dense concrete and must be taken into consideration in the design process.

The shear capacity of reinforced lightweight aggregate concrete is generally considered to be lower than that of normal weight concrete. Because the aggregate is weaker than the cement matrix the failure plane tends to go through the aggregate particles, leaving a relatively smooth surface. Thus the aggregate interlock, the mechanism by which a proportion of the shear is carried across the shear crack, is lower. The British Code for the design of bridges, BS 5400,[9] reduces the ultimate shear stress and the maximum shear stress to 80% of those permitted for normal weight concrete of the same grade, irrespective of the type of aggregate used. This is in line with BS 8110, the Code of Practice for structural concrete.[10] The American Concrete Institute Building Code[11] has a different approach, the shear capacity being taken as a function of the tensile capacity of the lightweight concrete which, for some aggregates, can be as high as that of normal dense concrete.

Because of uncertainties over the bond characteristics of lightweight concrete the stresses are reduced to 80% of those for deformed bars in dense concrete (50% for plain bars). This reduction is because of the possibility that, with a relatively weak aggregate, failure may be due to a cylindrical plug of concrete being pulled out, shearing along the tops of the ribs rather than crushing at the faces of the ribs. However, there is some evidence that the assumed type of failure does not occur with the higher grade aggregates, and hence this reduction may not be necessary. Further research is required in this area. The permissible bearing stresses inside bends are two-thirds of those allowed in dense concrete, again reflecting the lack of experimental data.

The elastic modulus of lightweight concrete is lower than that of dense concrete. For a given strength, the British code multiplies the dense concrete elastic modulus by $(D/2300)^2$, where D is the density of the lightweight concrete. Thus for structural grades of lightweight the elastic modulus is likely to be about 60% of that for dense concrete. This reduction will lead to increased deflections both in the short and long term. Neither should, however, lead to problems for the bridge designer that cannot be overcome. The lower elastic modulus will be of benefit in certain situations as it will lead to lower loads due to deformations, such as settlement or thermal strains. The latter will be further reduced by the low coefficient of thermal expansion: the Bridge Code, BS 5400,[9] recommends a figure which is about 60% that for normal concrete.

Because of the limited experience of the long term behaviour of lightweight aggregate concrete, the approach to durability is somewhat conservative: in the UK the Bridge Code requires 10 mm more cover to the reinforcement than the figure required for dense concrete of the same grade. However, Grimer[12] gives details of exposure tests on a range of lightweight aggregate concretes. He reported that no significant corrosion of the reinforcement had occurred with any of the concretes of structural grade after 6 years' exposure in an aggressive industrial environment. A survey of the use of lightweight aggregate concrete in bridges in the United States and Canada[13] has been carried out on approximately 2 km^2 of bridge decks, varying in age up to 39 years old. Out of the forty or so highway authorities taking part, only four suggested that there had been any problems while one commented that the lightweight was in fact better than normal concrete. Unfortunately, the types of aggregate used are not specified.

7.2.4 Design Considerations for Prestressed Concrete

The sections in the Bridge Code, BS 5400,[9] dealing with the design of prestressed members exclude the use of lightweight concrete. However, the Code for the structural use of concrete, BS 8110,[10] does allow lightweight but notes that the creep and shrinkage are likely to be higher than for normal weight concrete, leading to greater prestressing losses. The designer is referred to specialist literature: Spratt[1] suggests that for structural grades of lightweight the creep should be taken as 30% higher than for dense concrete and the shrinkage 40% higher. Section 5 of Part 2 of BS 8110 gives additional considerations in the use of lightweight aggregate, generally modifying the clauses for reinforced concrete. In general, the prestressed clauses could be modified along similar lines, but no specific guidance is given.

One area of concern may again be shear. In the Bridge Code the shear of sections uncracked in flexure, given in eqn (7.1) below, is based on the principal tensile stress being equal to the tensile capacity, f_t, of the concrete,

$$V_{co} = 0.67bh\sqrt{(f_t^2 + f_{cp}f_t)} \tag{7.1}$$

where f_{cp} is the compressive stress due to prestress at the centroidal axis, modified by a partial safety factor. For dense concrete, f_t is taken as $0.24\sqrt{f_{cu}}$. Applying a reduction factor 0.8 to the tensile strength gives an amended value of $f_t = 0.2\sqrt{f_{cu}}$. The design of sections cracked in flexure, given below, is based on an empirical equation, derived from tests on normal weight members.

$$V_{cr} = 0.037bd\sqrt{f_{cu}} + (M_{cr}/M)V \tag{7.2}$$

where

$$M_{cr} = (0.37\sqrt{f_{cu}} + f_{pt})I/Y \tag{7.3}$$

in which f_{pt} is the stress due to prestress, modified by a partial safety factor, at the extreme tensile fibre, and M and V are the moment and shear at the section under consideration.

While eqn (7.3), giving the cracking moment, M_{cr}, may be readily amended for lightweight by reducing the factor 0.37 to 0.3, it is unclear how eqn (7.2) should be modified, if at all.

When account is taken of the tensile capacity of the concrete, such as the permitted flexural tensile stresses under service conditions, the codes require that the values given for dense concrete should be multiplied by 0.8. Similarly, unless experimental evidence is available, transmission lengths in pretensioned members should be taken to be 25% greater than for dense concrete.

7.3 HIGH STRENGTH CONCRETE

7.3.1 Introduction

The word 'high' is, of course, a relative term. Twenty-five years ago $40\,\mathrm{N/mm^2}$ concrete was considered high strength, while nowadays it is commonplace in precast work. For the construction of the oil production platforms in the North Sea a $70\,\mathrm{N/mm^2}$ concrete is currently used, and a $100\,\mathrm{N/mm^2}$ concrete is being developed. In Japan, railway bridges have been built using $85\,\mathrm{N/mm^2}$ concrete for the last ten years. Such mixes require careful selection of the aggregates and their gradings and the use of admixtures to reduce the water content while retaining the workability of the fresh concrete. In addition, retarding admixtures may be required to reduce the heat of hydration and hence the risk of thermal cracking. Admixtures are discussed further in a later section. The recently published American Concrete Institute State of the Art Report on high strength concrete[14] suggests that a doubling in strength from 50 to $100\,\mathrm{N/mm^2}$ can be achieved at a cost increase of only 50%. In other words, the cost of carrying a given load is substantially less when the higher strength concrete is used.

The benefits are similar to those for lightweight concrete. To the cost savings resulting from the use of lower volumes of concrete must be added reduced shuttering costs and lighter foundations, In addition, the structure may be simpler: for example, Jobse and Moustafa[15] showed that increasing the design concrete strength, for the particular bridge they were considering, from 50 to $80\,\mathrm{N/mm^2}$ roughly halved the number of precast beams required. Fig. 7.2 shows an idealised cross-section of the bridge. The deck thickness had to be increased somewhat to take the increased transverse moments due to the increased beam spacing, but there was a net reduction in structural weight. In addition, they point out that the reduced number of bearings required would simplify maintenance of the bridge.

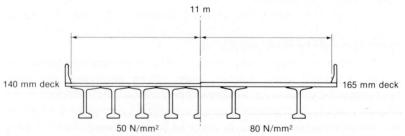

Fig. 7.2. Comparison of bridge cross-sections.

The ACI Report pays considerable attention to the use of high strength concrete in columns, where reduced cross-sections may be used to carry a given load or, alternatively, the amount of reinforcement may be reduced. In the former case there will be cost savings resulting from the reduced shuttering required, and in the latter the detailing will probably be simpler, with fewer splices or laps required in long columns. The reduced cross-section may, of course, mean that buckling may become critical in some situations.

7.3.2 Design Considerations for Reinforced Concrete

From the point of view of structural design, high strength concrete generally lies outside the provisions of codes of practice and therefore reference has to be made to specialist literature.

The ACI State of the Art Report[14] gives some general recommendations. For the ultimate moment of resistance of members in flexure, it is assumed that the concrete is either at a uniform stress of $0.4f_{cu}$ over the whole depth of the stress block or follows a parabolic–rectangular distribution with a maximum value of $0.45f_{cu}$. Both assumed stress distributions are approximations to the actual stress distribution at ultimate, but the shape assumed has little effect on the predicted bending moment. The important factors are the total compressive force and the lever arm. The ACI report suggests that the assumption of a rectangular block can be used with equal confidence for concretes up to at least $100 \, \text{N/mm}^2$ strength.

In British codes the limiting compressive strain is taken as 0.0035 at the extreme fibre. The Report[14] gives experimental strains for a range of concrete strengths. Although there is a considerable scatter, depending on the section geometry and other factors, the assumed value of 0.0035 is a reasonable mean throughout the range of concrete strengths, up to about $100 \, \text{N/mm}^2$. Thus, for under-reinforced members, the bending capacity for high strength concrete may be determined using the same approach as for normal concrete.

Test evidence suggests that high strength concrete may be more brittle than normal weight concrete. This is illustrated in Fig. 7.3, from which it may be seen that the falling branches of the stress–strain curves are almost non-existent above about 60 grade. With a sufficiently stiff testing machine, however, a falling branch can be detected but the curve still decreases at a much greater rate than for lower-strength concretes. Although the safety factors incorporated in the approach for determining the flexural capacity will adequately cover these differences, it may be necessary to consider the ductility requirements and the minimum steel requirements.

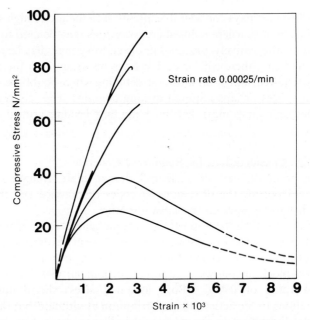

Fig. 7.3. Compressive stress–strain curves for concretes of different strengths.

For reinforced concrete the British Bridge Code[9] requires a strain of 0·0035 in the concrete and that the reinforcement is yielding. As an alternative, the steel strain may be less if the calculated strain in the concrete, due to the application of 1.15 times the ultimate load, does not exceed 0·0035. These requirements should be equally valid for high strength concretes.

The minimum areas of tension reinforcement are based only on the grade of steel being used (0·15% for 460 and 0·25% for 250, based on the effective depth of the section). The principle behind the adoption of these values is that the tensile capacity of the steel should exceed the tensile capacity of the concrete when it cracks. Thus the minimum area provided should increase with increasing tensile capacity, i.e. with the square root of the cube strength. On the basis that 40 N/mm^2 is the maximum grade quoted for reinforced concrete, it is suggested that the quoted minimum area of tension reinforcement should be multiplied by $\sqrt{}$(concrete grade/40) when high strength concrete is used.

It is suggested in the ACI Report that the shear capacity of reinforced beams made with high strength concrete does not increase indefinitely with

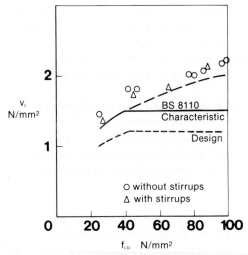

Fig. 7.4.   Shear capacity of high strength concrete beams.

concrete strength, because aggregate interlock accounts for a lower proportion of the section strength. This may justify the approach taken by the British Code which makes no allowance for the increased shear capacity of beams or slabs made with concretes above 40 N/mm². Not surprisingly, there is little test evidence for concretes of higher strengths, but some work has been carried out by Mphonde and Frantz at the University of Connecticut.[16] Analysis of their results shows a steady gain in shear strength with increasing concrete strength. This is illustrated in Fig. 7.4, which compares the results with the strengths predicted by the BS 8110 equation both for beams with and without shear reinforcement and suggests that the 40 N/mm² cut-off is not justified. The Code takes the maximum shear stress as being proportional to the square root of the cube strength up to 60 N/mm². The approach should be equally valid for higher strengths.

The Bridge Code requires a minimum area of shear reinforcement in beams equivalent to a shear stress of 0·4 N/mm². This figure is deemed to be reasonable for 'normal' concrete strengths but its derivation is obscure; it may be sensible to work to a slightly higher figure for higher strength concretes. At sections at which the applied shear stress exceeds the shear capacity of the concrete, reinforcement must be provided to carry the excess plus the same 0·4 N/mm². This additional requirement, which is not included in BS 8110, reflects uncertainty over the concrete contribution in members that are subjected to repeated loadings. Limited test evidence would suggest that, for interface shear at least, the long-term capacity is roughly half that of the short-term capacity, and hence it may be reasonable to replace the figure of 0·4 N/mm² by half the value of the ultimate shear stress for the concrete given in Table 8 in BS 5400.

7.3.3 Design Considerations for Prestressed Concrete

The provisions for prestressed concrete in the Bridge Code extend to higher grades of concrete than those for reinforced (60 as against 40), and thus the limitations on the use of high strength concrete are likely to be fewer. At the ultimate limit state the strength is assessed in exactly the same way as for reinforced concrete and will be equally applicable, as discussed earlier. Similarly, the ductility requirements should still be valid.

As discussed in the section on lightweight concrete, the British Code gives expressions for the shear capacity of sections that are uncracked in flexure and for sections cracked in flexure, the expressions being valid for concretes up to 60 N/mm². The ACI Report states that the generally accepted assumption of the splitting tensile strength being related to the square root of the compressive strength is valid for concretes up to 100 N/mm². This is illustrated in Fig. 7.5, based on data from Carrasquillo

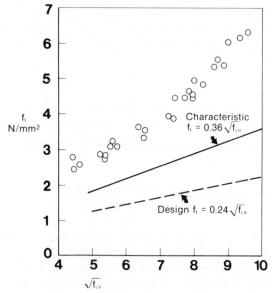

Fig. 7.5.   Tensile strength of concrete (split cylinder).

et al.,[17] which shows that the design value for the tensile strength used when determining the shear of the uncracked section $(0.24\sqrt{f_{cu}})$ is conservative for both normal and high strength concretes. Hence the approach used for sections that are uncracked in flexure, eqn (7.1), may be used throughout the range. The expression for the shear of sections cracked in flexure, eqn (7.2), is empirical, based on tests on beams of normal strength. The expression for the cracking moment contains a value of $0.37\sqrt{f_{cu}}$ for the modulus of rupture; Fig. 7.6 shows that this is conservative for concretes up to $100\,\text{N/mm}^2$. The second part of eqn (7.2) dominates the shear capacity and hence it may be assumed that the approach is valid for all concrete strengths. The provision of shear reinforcement is the same for both reinforced and prestressed sections and hence the remarks in the previous section apply equally here.

The allowable tensile stresses in the concrete under service conditions are given in terms of the square root of the cube strength. In the light of the previous paragraph they should be equally valid for all grades of concrete.

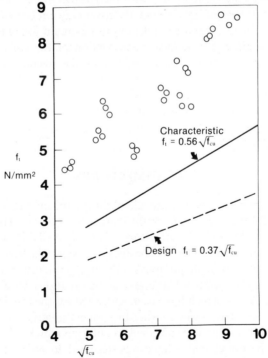

Fig. 7.6.   Tensile strength of concrete (modulus of rupture).

Creep plays a significant role in the losses occurring in prestressed members. The creep taking place at a given level of stress is, broadly speaking, inversely proportional to the strength of the concrete at the time the load is applied. (There are data available[18] for concretes up to $60\,N/mm^2$ but the British Bridge Code makes no allowance for strengths above $40\,N/mm^2$ at the time of stressing.) Thus, for a given level of prestress, the creep in a high strength concrete member will be less than that in the equivalent normal strength member. The ACI State of the Art Report[14] suggests that the shrinkage of high strength concrete will be similar to or lower than that of normal strength concrete. As losses will be lower, it may be possible to reduce the number of prestressing strands provided.

One possible drawback to the use of very high strength concrete, either reinforced or prestressed, may be that the requirement for the provision of adequate cover to the reinforcement may limit the extent to which section sizes may be reduced. From the point of view of durability, the cover requirements in the Code reduce with increasing concrete grade, up to a maximum of $50\,N/mm^2$. In addition, the cover may not be less than the size of the bar, to ensure adequate bond, nor the maximum size of the aggregate plus $5\,mm$ to allow for adequate compaction of the concrete. Finally, considerations of lateral stability will limit the overall dimensions of members. The British Bridge Code suggests a limit to the distance between lateral restraints of $60b$ or $250b^2/d$, whichever is the lesser. For a cantilever the limits are $25b$ and $100b^2/d$. In these expressions b is the breadth of the compression face of the beam midway between restraints and d is the effective depth.

7.4 ADMIXTURES

Admixtures are materials that are added during the mixing to modify the properties of either the fresh or the hardened concrete. Higgins[19] has reviewed the wide range of admixtures that are available. Two that are likely to be of significance for bridges are superplasticisers and air-entraining agents, though pumping aids and accelerating or retarding admixtures may be used to advantage in some situations. He also mentions corrosion inhibitors, which have been suggested for the last 25 years or so as being a viable solution when a waterproof membrane is, for some reason, not possible for a structure subjected to chloride attack.

It should be noted that admixtures designed to alter one particular property of the fresh or hardened concrete may also have secondary effects

Fig. 7.7.   Flowing concrete.

which may possibly be detrimental. For example air entrainment, as discussed below, improves frost resistance but lowers the compressive strength. Hence care must be taken in selecting a suitable admixture. Commercial admixtures may contain materials that separately would perform two different functions. Care must also be taken in the use of admixtures to ensure that correct proportions are used: American Concrete Institute Committee 212 has given detailed guidance.[20]

7.4.1 Superplasticisers

One modification to conventional concrete that has had a significant effect on site practice, particularly in the USA, is the availability of flowing concretes. These are concretes that can be placed with the minimum of effort. Flowing concrete (Fig. 7.7) can be made by adding an excess of water but the resultant concrete will be porous and prone to rapid carbonation or the ingress of aggressive chemicals. The strength of the concrete will be low and the concrete will be liable to shrink and crack.

Superplasticisers provide one way to increase the flowability of concrete without these detrimental effects.[21] They can be used in two ways:

(a)   to reduce the water content in a mix, without affecting the workability; or

(b)   to make a stiff concrete very workable.

In the latter case it is possible to design a stiff mix which will give the required final properties and then add a superplasticiser to the mix so that it can be pumped and placed with the minimum of effort, without segregation or bleeding.

The use · of superplasticisers enables concreting to be carried out efficiently in places where the reinforcement is extremely congested, for example round the anchorages at the ends of prestressed members. Superplasticisers act only in the short term, for a maximum of 3 to 4 hours for the latest formulations. The mechanical properties of the hardened concrete will be very similar to those of the untreated concrete. In the long term superplasticisers have no direct effect and the indications, after about 30 years' experience world-wide, are that the durability of superplasticised concrete is as good as that of untreated concrete. In fact it is likely to be better because of the higher degree of compaction obtained.

7.4.2 Air Entraining Agents

The entrainment of controlled quantities of air in the concrete generally improves its resistance to frost damage. Air-entraining admixtures are used to give about 5% of air, in the form of small bubbles about 0·5 mm in diameter and evenly distributed through the concrete. When concrete freezes, the moisture trapped in it turns to ice, expanding as it does so, which can cause damage. With entrained air the ice expands freely into the voids and no damage results.

One disadvantage of air entrainment is that it slightly reduces the strength of the concrete. Higgins[19] suggests a reduction of about 15%, but points out that air entrainment has a plasticising effect on the mix which means that some reduction of the water content, and consequent increase in the strength, may be possible.

7.4.3 Pumping aids

In recent years pumping has become a common way of placing concrete. Not all mixes are suitable for pumping: some can be improved by use of a pumping aid admixture which thickens the cement paste and makes the mix less liable to bleeding. Some of the admixtures cause air to be entrained in the mix, and therefore a defoaming agent may be required as well. In addition, many of the admixtures cause some retardation of the setting of the concrete.

7.4.4 Accelerators

Accelerators are used to shorten the setting time of the mix and to increase the rate of gain of early-age strength. They can be beneficial in precast work, enabling a speedy turn-round of shuttering, and can also be of use for *in situ* concrete work, allowing props to be removed earlier. In addition, a high early strength will be important for concrete that is liable to damage by freezing. In the past the most common accelerator used was calcium chloride, but this has led in the long term to increased corrosion of embedded reinforcement and hence the chloride content of permitted accelerators is now limited; BS 8110[10] sets a limit of 2% by weight of the admixture or 0·03% by weight of the cement.

7.4.5 Corrosion inhibitors

It has been suggested that certain admixtures can be used to inhibit corrosion of the reinforcement in the presence of chlorides.[20] One that shows promise is calcium nitrite, which has been used in bridges in certain parts of the United States since 1979.[22] When corrosion takes place in untreated concrete, the ferrous ions at the anode pass into solution and, in a secondary reaction, are converted to rust. With the calcium nitrite, ferric ions are formed which are insoluble and hence stay on the surface of the reinforcement, preventing further corrosion. Accelerated corrosion tests appear to show that the age at which significant corrosion occurs is considerably increased by the additon of 2% by weight of calcium nitrite to the cement. However, there would not appear to be data available on longer-term behaviour, nor on actual in-service use. It is not clear whether calcium nitrite will have any effect on the rate of corrosion once it is initiated. It is reported also that concrete treated with calcium nitrite has a slightly higher compressive strength than untreated concrete.

7.5   CEMENT REPLACEMENTS

There are several materials that may be added to cement which have pozzolanic properties, that is they react with the lime that is produced during the hardening of Portland cement. Naturally the resulting concrete has somewhat different properties which should be considered before they are adopted.

7.5.1 Pulverised Fuel Ash and Granulated Blast-furnace Slag

Pulverised fuel ash, generally known as PFA, is the fine ash produced when powdered coal is burned in power stations. It is obtained from the electrostatic precipitators used to clean the flue gases and hence minimise atmospheric pollution. Blast-furnace slag is the waste from the smelting of iron, which is rapidly cooled by water to form a granulated material which is later ground to form a powder, generally known as GGBFS. When mixed with Portland cement, both materials have similar properties which are outlined below. Blending of the Portland cement and the replacement may either be carried out by the manufacturer or on site. In the first situation the product will be required to comply with the necessary specifications or standards. In the second, while the cement and the replacement will individually comply, problems may arise with the proportioning on site.

The strengths quoted in codes of practice are those at 28 days. Blended cements have a slower rate of gain of strength than ordinary Portland cements (OPC) at early ages but a higher gain at later ages. In other words, for a given specified 28-day strength, the strength of the concrete during the life of the structure will be higher if cement replacement is used. However, the construction programme may be delayed because the slower rate of strength gain will lead to delays in removing formwork and props, and a longer period before prestressing can take place. This can be particularly significant when concrete is placed during the winter. Longer curing periods are required when blended cements are used: BS 8110[10] recommends about a 50% increase in the period required under a given set of conditions. On the other hand, the rate at which heat is generated during hydration is lower with blended cements, which may reduce the problems with large pours.

It is commonly assumed that using a blended cement does not alter the mechanical properties of the hardened concrete, i.e. the elasticity, creep etc., assuming that the strength is unaltered. It is also likely that the abrasion resistance of blended cement concrete which has been well cured will be similar to that of OPC concrete.

The factors that influence the age at which corrosion of reinforcing steel may start are carbonation of the concrete and, in a bridge, the penetration of chlorides from de-icing salts. It is generally accepted that the depth of carbonation in a member is a function of the concrete strength and hence it is likely that the behaviour of well cured OPC and blended cement concretes will be similar. However, pozzolans combine with much of the calcium hydroxide which provides the reserve of alkalinity to protect the steel. There is evidence that the penetration of chlorides into blended cement concretes is slower than in OPC concretes. Once corrosion has started, the rate is governed by the permeability of the concrete cover. Because the pore size is smaller in blended cement concretes, the permeability will be lower and hence the rate of corrosion would be expected to be lower. However there would not appear to be quantitative evidence to support this, though the behaviour with slag may be better than that with PFA.

The resistance to freeze–thaw attack would appear to be similar for a given grade of concrete but, because of the slower initial gain in strength, young concretes made with blended cements may be more susceptible.

Blended cements may help in preventing alkali–silica reaction, which occurs when alkalis in the cement react in the presence of water with certain forms of silica in the aggregate to form a gel which can exert a sufficient pressure to damage the concrete. When the aggregate contains more than a critical amount of reactive silica, the alkali content of the cementitious material must be kept low: a level of 0·6% is recommended by BS 8110.[10] One way of achieving this is to use blended cements, containing at least 30% of PFA or 50% of GGBFS, although it has been suggested that not all pulverised fuel ashes may be suitable for this purpose, since they contain alkalis themselves.

7.5.2 Silica Fume

Silica fume, otherwise known as microsilica, consists of the very fine particles of amorphous silica which are collected in the dust removal systems during the manufacture of silicon and ferro-silicon metals. It has been used for a number of years in Scandinavia and North America, initially as a cement replacement but later as an additive to give improved properties. Silica fume is an efficient pozzolan, 1 kg replacing 3 to 4 kg of cement with no change in the concrete strength. However, because silica fume is extremely fine, with a specific surface area in the region of 30 times that of normal Portland cement, the water demand is very high. Thus it is generally necessary to use a superplasticiser in the mix to give a reasonable level of workability.

The addition of small quantities of silica fume to good quality mixes leads to appreciably higher strengths, due to the very fine particles packing into the small voids that normally exist. Malhotra and Carette[23] report concrete strengths in excess of $200 \, \text{N/mm}^2$ and mortar strengths approaching $300 \, \text{N/mm}^2$. The densities are slightly higher than would be expected for normal concretes. Although there is some suggestion that the early-age strength gain is slower than for normal concretes, the authors indicate a substantial gain after 28 days.

There would appear to be little information on the in-service durability of reinforced concrete incorporating silica fume, though Hjorth[24] suggests that the performance will be similar to or better than that of normal concrete. Carbonation rates are similar but the rates for chloride penetration are lower, which means that the time for corrosion of the reinforcement to start will be longer in silica fume concrete. Once corrosion is initiated, the lower rate of diffusion of oxygen should lead to a slower reaction. To date, in fact, the use of silica fume concrete has been largely limited to the chemical industry where it has been adopted in areas subjected to severe chemical attack or to abrasion. It may be that the most likely application for bridges will be in the form of an overlay (a trial length was laid in Ohio in 1984) or else as permanent formwork, which is discussed later.

Silica fume has been shown to be highly effective in the control of alkali–silica reaction. In Iceland, where the Portland cements have a high alkali content and the aggregates are reactive, 7·5% of silica fume is currently added to all Portland cement produced.

7.6 POLYMER MODIFIED CONCRETES

Polymer modified concretes are those in which polymers are added to conventional concrete, either during the mixing process or by impregnation of the mature concrete. Polymer concretes are those which consist solely of a polymer and aggregate, without any Portland cement, and are not considered here. The addition of polymers will lead to improved mechanical properties:[25] Table 7.1 gives some typical values. In addition, the durability of the concrete and its ability to prevent corrosion of the reinforcement are both improved.

Polymer impregnation is carried out by drying the concrete to the desired depth, allowing the polymer in a monomeric form to soak into the concrete and then polymerising the monomer by raising the temperature. The majority of the experience of the application of the process to bridges has been the repair of deteriorated decks. However, polymer impregnation was

TABLE 7.1
Typical Properties of Polymer Modified Concretes[25]

	Portland cement concrete	Polymer impregnated concrete	Polymer cement concrete
Compressive strength (N/mm²)	35	140	40
Tensile strength (N/mm²)	2·5	10	5·5
Young's modulus (kN/mm²)	24	41	14

used on the tops of the precast tee-section units for the Santa Rosa Bridge in Washington in 1977, the treatment taking place at the precasting factory. In 1981 it was estimated that approximately 50 000 bridges in the USA had serious corrosion of the reinforcement in the decks, largely due to the use of de-icing salts.

Current rehabilitation methods generally involve the removal of the chloride-contaminated concrete and its replacement. Polymer impregnation would appear to offer an attractive alternative solution by sealing the pores in the existing concrete and restricting the ingress of moisture and oxygen. Experience in Pennsylvania[26] has shown that the process not only prevents corrosion but also arrests already active corrosion. In addition, the resulting concrete is more resistant to wear. However, the study did not show the improvement in mechanical properties mentioned earlier, probably due to the presence of small cracks already existing in the concrete. Weyers and Cady[27] have reported on a method of grooving the concrete to speed the impregnation process. They estimate that the cost of polymer impregnation would be about 60% of that of replacing the surface of the deck.

While polymers have been used chiefly to improve the durability of concrete and to provide a suitable top surface for bridges, the improved material properties listed in Table 7.1 will lead to some overall improvement in the member strength. However, very little work has been carried out on this aspect. Dikeou[28] suggests that, because the material is brittle, an elastic approach to predicting the ultimate flexural capacity may be more appropriate than the usual assumptions of plastic behaviour. However, any reinforcement incorporated in the concrete will tend to make the behaviour more ductile. Short-term deflections will be reduced because of the increased elastic modulus, and creep is also significantly reduced. But impregnation of the complete member is unlikely to be practicable. In addition, the improvement in properties that can be achieved depends on the initial quality of the concrete: in high quality concrete the porosity is low and hence the amount of polymer that can be introduced is low. However, in certain cases partial impregnation of a member might prove to be an economical solution, with polymers only being used in areas where they can be of particular use. For example, Dikeou[28] has reported that polymer impregnation increases the shear capacity of beams without shear reinforcement by about 60%, and that anchorage stresses are increased. This could be advantageous at the ends of pretensioned beams, reducing the transmission length and the amount of shear reinforcement that needs to be provided.

7.7   FIBRE REINFORCED CONCRETE

Concrete has a low tensile capacity. In conventional reinforced concrete steel reinforcing bars are used to carry any tensile forces and in prestressed concrete the member is preloaded, before the external loads are applied, which raises its effective tensile capacity. In both cases an adequate thickness of concrete must be provided round the steel to prevent corrosion. For many years attempts have been made to improve the tensile capacity of the concrete by the inclusion of fibres and hence avoid the necessity for including steel reinforcing bars or prestressing cables. A homogeneous material would lead to simpler structural members and ones that may be thinner because the necessity to provide adequate cover to conventional reinforcement would no longer exist.

Research has been carried out on a variety of fibres over the years; in fact the first patent was taken out over 100 years ago.[29] The fibres that are of most interest for use in bridges include steel, glass and polymers such as polypropylene. Steel fibres are available in a range of geometries (i.e. straight, crimped, etc.) and should be stainless or galvanised to prevent rusting of those exposed at the surface of the concrete. Glass fibres have a high modulus of elasticity and strength. However, the resulting concretes tend to become brittle with age and hence should not be used for load-bearing structural elements. Polymer fibres have significantly lower modulus of elasticity than steel, but are extremely durable. The bond between the fibre and the concrete is poor and hence mechanical anchorage, for example by twisting the fibres, is generally required.

The properties of fibre reinforced concrete are influenced by the type and amount[30] of fibre added. For this reason the ACI State of the Art Report[30] suggests that mixes should always be tested before use to check the assumed properties. However, Table 7.2, taken from ref. 31, gives an indication of the properties that can be obtained.

The mixing of fibres into the concrete must be carried out with care. The ACI Guide for using steel fibres[32] recommends that the mortar on concrete be fully mixed and then the fibres added: this helps to avoid the problem of the fibres forming into tight balls during the mixing process. The placing and finishing of fibre concrete is carried out in the same manner as for conventional concrete, though care must be taken that the concrete is not overworked during finishing.

It is possible that fibres could be used to replace some of the conventional reinforcement in structural members. For example, the amount of shear reinforcement could be reduced,[33] which could be particularly useful near

TABLE 7.2
Typical Properties of Fibre Cement and Concrete[31]

Property	Steel fibres	Glass fibres	Polymer fibres
Elastic modulus (N/mm² × 10³)	Controlled primarily by the mortar or concrete modulus. Fibres may entrap some air and hence reduce the modulus		
	10–20	30	5–10
Compressive strength	Controlled primarily by the mortar or concrete properties until the matrix cracks. Subsequent deformation is ductile and controlled by fibre properties, fibre–matrix bond and pull-out forces.		
Tensile strength (N/mm²)	1·5% fibres by volume	5% fibres by weight	Polypropylene 7% by volume in mortar
	mortar 2–4·5	28 days 15	2–3
	concrete 2–4	2 years* 13	
Flexural strength (N/mm²)	1·5% fibres by volume 6–8	28 days 38	Polypropylene 7% by volume in mortar 5–6
	3% fibres by volume 8–12	2 years* 36	
Impact strength (kJ/m)	No agreed test methods Generally, impact resistance is 1000–3000% of equivalent plain concrete	Izod 28 days 21 2 years* 20	No agreed test methods Generally, impact resistance is 200–1000% of equivalent mortar or concrete matrix
Creep and shrinkage	Controlled primarily by the mortar or concrete properties		

* Natural weathering.

the ends of prestressed members, reducing congestion. Fig. 7.8 illustrates a situation in which fibres have been used to increase the concrete contribution to the shear capacity by 100%, and the number of stirrups has been reduced to about one-third. At present the cost is likely to be too high for fibres to be used throughout a member, apart from exceptional circumstances, for example where conventional reinforcement would require complex shuttering.

The most likely application at present for steel fibre concrete in bridges is in the form of overlays to the deck to give a highly durable running surface

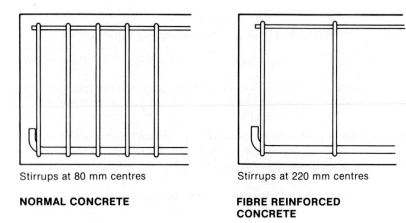

Stirrups at 80 mm centres   Stirrups at 220 mm centres

NORMAL CONCRETE   **FIBRE REINFORCED CONCRETE**

Fig. 7.8.   Shear reinforcement of beams incorporating fibres.

and to provide good protection to the structure beneath. Melamed[34] reported that a 65 mm fibre reinforced overlay used on some bridges in Alberta was 25% cheaper than a conventional 100 mm overlay, and preliminary indications were that the performance was excellent.

Glass reinforced cement is widely used for the production of thin sheet products. Currently GRC panels are used as permanent formwork to speed the construction process.[35] The panels are durable and impermeable, with good freeze–thaw resistance. Hence they could be used to give added protection to the reinforcement in concrete and could possibly be used for remedial work where the cover provided has been inadequate. This will be discussed later.

7.8   PERMANENT FORMWORK

The use of permanent formwork has obvious advantages, including a reduction of the labour and material costs involved with conventional formwork which must be removed after the concrete is cast. Provided the formwork is sufficiently robust, no additional supports are required, leaving the area under the structure clear. Fig. 7.9 shows permanent formwork panels prior to use and installed in a bridge before the *in-situ* concrete was cast. In the past, the chief consideration for the design of permanent formwork has been that it should be capable of carrying the weight of the wet concrete and the associated construction loads without

(a)

(b)

Fig. 7.9.   Permanent formwork: (a) before installation and (b) in place.

undue deflection. In areas where it is exposed, the material must be durable and have a suitable surface finish. However, this approach is rather limited. The choice of a suitable material for the permanent formwork can provide additional long-term protection for the concrete. In addition, the formwork, acting compositely with the concrete, will increase the load-carrying capacity. The former consideration could be particularly significant for bridge structures that are subjected to de-icing salts.

Glass fibre reinforced cement (GRC) has been used to make permanent formwork for the last 20 years. A wide range of profiles is now available[35] from simple flat sheets suitable for spanning 500 mm or so up to a complex ribbed element suitable for a span of nearly 3 m. These figures are for a typical *in-situ* bridge deck 250 mm thick. As GRC permanent formwork forms a dense, relatively impermeable skin to the concrete, the passage of water and chlorides to the steel is significantly reduced, improving the durability. This has been acknowledged by the Norwegian Roads Directorate, who have recommended that when GRC permanent formwork is used in bridge deck construction, the main reinforcement needs only a further 10 mm of concrete cover in addition to the GRC skin.[35] Norwegian practice is to use corrugated panels, the tops of the corrugations acting as spacers for the distribution steel which is laid directly onto them. It is not clear whether the additional 10 mm is required from the point of view of durability or to provide sufficient concrete under the main reinforcing bars to ensure adequate bond.

The piers and abutments of bridges are sprayed with salt-laden water during the winter. Here again GRC could provide a dense impermeable skin to the concrete, either being used as permanent formwork or fixed to the surface of conventionally cast concrete. Further savings could be made as a result of being able to use a lower grade of concrete for the member. Reducing the cover requirement would have the additional advantage that member sizes could be reduced slightly, with a consequent reduction in weight. Alternatively, the load capacity for a given member depth could be increased slightly. Fig. 7.10 shows a 6% increase in flexural capacity due to the increased effective depth resulting from the use of permanent formwork for a 400 mm deep beam.

While the strength of the permanent formwork is required simply to support the *in-situ* concrete until it has matured sufficiently to support itself, more economic structures would result if account was taken of the resulting composite action. Profiled steel sheeting is widely used as permanent formwork in buildings, forming the tensile reinforcement of the resulting composite slab. Galvanised sheeting, to avoid problems with

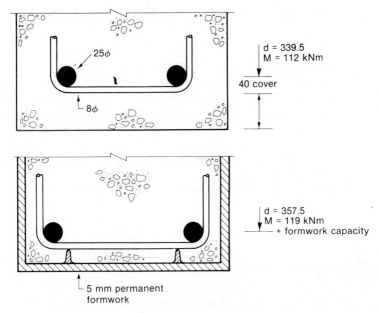

Fig. 7.10.   Increased beam capacity resulting from the use of permanent formwork.

corrosion, has been used in the USA. Ferrocement would appear to be a suitable alternative. It is not a new material and has, in fact, been used for over a hundred years, but not in bridges. It is formed by plastering a cement-rich mortar onto a mesh of small diameter wires or expanded metal to form a very strong, impact-resistant material. In the West its most common use has been in the construction of yacht hulls, though in the developing countries the technique has been used to build a wide variety of storage structures. Complex shapes can be fabricated which would be capable of spanning wider gaps and carrying greater loads than the present profiled steel sheets. Work is currently under way at the Israel Institute of Technology[36] on the bending behaviour of composite beams made with ferrocement permanent shuttering both for the base and sides and a low strength concrete filling. Problems were experienced during the early tests due to breakdown of the bond between the ferrocement and the concrete at high loads, but even so, the composite units carried 25% more load than a simple beam of the same dimensions. Further work is going ahead on a range of L- and Z-shaped units that can be used to build up a number of different profiles.

7.9 NON-CORRODING REINFORCEMENT

The protection of the reinforcement to prevent corrosion is a significant consideration, particularly when the environment is corrosive, such as on bridge structures subjected to de-icing salts. This leads to the requirement for a thick concrete cover to ensure adequate durability. Some of the preceding sections have dealt with ways of improving the quality of the cover and hence reducing the thickness required. Cathodic protection, the technique by which the electrical potential of the steel is increased to a level at which corrosion cannot take place, is widely used for both steel and concrete offshore structures.

Two different methods are employed: an impressed current, and the use of sacrificial anodes. In the first the structure is connected to the negative terminal of a DC power source, ideally using an anode which does not corrode. In the second the reinforcement is connected to anodes with a more negative corrosion potential than steel, such as zinc or aluminium. The current is reversed and corrosion now takes place at the anode, which is gradually used up. In both cases electrical continuity of the reinforcement is required. Cathodic protection has been applied to a number of bridges in the USA and Canada,[37] using both techniques, generally with a view to halting corrosion that was already taking place, but with limited success. The cost of running the protection is reported to be low in relation to the total maintenance budget.

There is a growing interest in the use of non-corroding reinforcement[38] as an alternative to providing cathodic protection. Stainless steel has been used for certain applications, but the cost is roughly ten times that of normal steel, which prevents widespread usage. Galvanising the steel is a less expensive option, doubling the basic price of the reinforcement, but is less satisfactory and can lead to problems when bars are bent. Damage must be made good with care, or localised corrosion may take place. In the USA, epoxy coated reinforcement is widely used in bridge decks. The coating, a layer 0·13 to 0·3 mm thick, has been found to prevent corrosion of the bar, though localised damage to the coating can lead to severe local corrosion. On the basis of American experience, the cost of using epoxy coated reinforcement is likely to be in the region of 10% more than that of using uncoated steel, which will have an insignificant effect on the total cost of the structure.

An alternative is to look at man-made fibres as a source of reinforcement. The recent development of long filaments that can be formed into ropes, or into parallel bundles encased in a sheath, have found applications in other branches of civil engineering such as soil stabilisation and reinforced earth. In the past their modulus has been much lower than that of steel, but recent developments have led to fibres with moduli of the order of 60% that of steel with a breaking load in excess of that of prestressing strand.[39] The fibres are extremely inert chemically and have excellent resistance to the corrosive action of salt. Apart from a reduction in the bond and anchorage, there should be no serious problems with using the material in reinforced concrete. Deflections will be higher than for conventional reinforcement, but these can be designed out by pre-cambering. Crack widths will be larger, but the only requirement for their limitation will be from the point of view of aesthetics, because concern over the effect of cracks on corrosion will have been eliminated.

Research is currently being undertaken on their possible use in prestressed concrete.[40] The fatigue properties are reported to be excellent. The paper suggests that the bundled filaments would be suitable for use as external tendons in bridges.

7.10   CONCLUSIONS

This chapter has attempted to give design engineers an indication of the very wide range of cement-based structural materials that are available. No specific solutions have been presented nor have any general recommendations been made, because the design of each individual structure must be carried out with due regard for all the factors involved. There will be many situations in which modified concretes have a part to play, though it may be appropriate to consider their use only for critical areas. The resulting structure should then be a more efficient, more economical solution, behaving satisfactorily throughout its design life.

REFERENCES

1. Spratt, B. H., *The structural use of lightweight aggregate concrete*, Cement & Concrete Association, London, Publication 45.023, 68 pp. (1974).
2. Concrete Society, *Design and cost studies of lightweight concrete highway bridges*, Special Publication, London, 16 pp. (March 1986).
3. Concrete Society, *A case study of the comparative costs of a building constructed using lightweight aggregate and dense aggregate concrete*, Technical Paper 106, London, 20 pp. (1983).

4. Concrete Society, *A review of the international use of lightweight concrete in highway bridges*. Technical Report 20, London, 15 pp. (1981).
5. Kerensky, O. A., Robinson, J. and Smith, B. L., The design and construction of Friarton Bridge, *The Structural Engineer*, **58A**(12), 395–404 (Dec. 1980).
6. Goodfellow, R. G. and Fordyce, M. W. Cable-stayed solution for Ealing Broadway Centre, *Concrete*, 11–12 (Jan. 1985).
7. Faessel, P., Teyssandies, J. P. and Virlogeux, M. Conception et construction du pont d'Ottmarsheim, *Annales de l'Institut Technique du Bâtiment et des Travaux Publics*, No. 391 (Feb. 1981).
8. Lutz, J. G. and Scalia, D. J., Deck widening and replacement of Woodrow Wilson Memorial Bridge, *PCI J.*, **29**(3) (May/June 1984).
9. British Standards Institution, BS 5400: 1984, *Steel, concrete and composite bridges, Part 4, Code of practice for design of concrete bridges*, London.
10. British Standards Institution, BS 8110: 1985, *Structural use of concrete, Part 1, Code of practice for design and construction, Part 2, Complementary recommendations for special circumstances*, London.
11. American Concrete Institute, Building Code requirements for reinforced concrete, ACI 318-83 (1983).
12. Grimer, F. J. The durability of steel embedded in lightweight concrete, *Concrete*, 125–131 (April 1967).
13. Transportation Research Board, *A questionnaire survey on uses of lightweight aggregate in concrete highway construction*, Washington, DC, Information series: Group 2: Nr. 7, 6 pp. (1977).
14. ACI Committee 363, State-of-the-art report on high strength concrete, *ACI J.*, **81**(4), 364–411 (July–Aug. 1984).
15. Jobse, H. J. and Moustafa, S. E. Applications of high strength concrete for highway bridges, *PCI J.*, 44–73 (May–June 1984).
16. Mphonde, A. G. and Frantz, G. C. *Shear strength of high strength reinforced concrete beams*, University of Connecticut Report CE 84-157 (June 1984).
17. Carrasquillo, R. L., Nilson, A. H. and Slate, F. O., Properties of high strength concrete subjected to short-term loads. *ACI J.*, **78**(3), (May–June 1981).
18. Gerstle, K. H., Material behaviour under various types of loading. High Strength Concrete, *Proc. Workshop*, University of Illinois at Chicago Circle, 43–78 (Dec. 1979).
19. Higgins, D. D., *Admixtures for concrete*, Cement and Concrete Association, Slough, Publication 45.041, 8 pp. (1984).
20. ACI Committee 212, Admixtures for concrete and Guide for use of admixtures in concrete, *Concrete International*, 24–65 (May 1981).
21. Hewlett, P. C. and Edmeades, R., Superplasticised concrete, Concrete Society Current Practice Sheet No. 94, *Concrete*, 31–2 (April 1984).
22. Walitt, A. L., Calcium nitrite offers long-term corrosion prevention, *Concrete Construction*, 356–61 (April 1985).
23. Malhotra, V. M. and Carette, G. G., Silica fume concrete—properties, applications and limitations, *Concrete International*, 40–6 (May 1983).
24. Hjorth, L., Microsilica in concrete, Nordic Concrete Research Publication No. 1, Paper 9, Nordic Concrete Federation, Oslo, 1–18 (1982).
25. American Concrete Institute, Applications of polymer concrete, SP-69, 222 pp. (1981).

26. Cady, P. D. *et al.*, *Evaluation of nine years' service with a deep polymer impregnation of a bridge deck*, PTI 8412, University Park, Pennsylvania State University, 31 pp. (1984).
27. Weyers, R. E. and Cady, P. D., Deep grooving—a method for impregnating concrete bridge decks, *Transportation Research Record 962*, Transport Research Board, National Research Council, Washington, DC, 14–21 (1984).
28. Dikeou, J. T., Development and use of polymer concrete and polymer impregnated concrete, *Progress in Concrete Technology* (Ed. Malhotra, V. M.), Canadian Centre for Mineral & Energy Technology, Canadian Government Publishing Centre, Hull, Quebec, 539–82 (1980).
29. Naaman, A. E., Fibre reinforcement for concrete, *Concrete International, Design and Construction*, 7(3), 21–5 (March 1985).
30. American Concrete Institute, *State of the Art Report on fiber reinforced concrete*, ACI 544, IR-82 (1982).
31. Clarke, J. L. and Pomeroy, C. D., Concrete opportunities for the structural engineer, *The Structural Engineer*, 63A(2), 45–53 (Feb. 1985).
32. ACI Committee 544, Guide for specifying, mixing, placing, and finishing steel fiber reinforced concrete, *J. ACI J.*, 81(2), 140–8 (March–April 1984).
33. Swami, R. N. and Bahia, H. M., The effectiveness of steel fibres as shear reinforcement, *Concrete International*, 7(3), 35–40 (March 1985).
34. Melamed, A., Fibre reinforced concrete in Alberta, *Concrete International*, 7(3), 47–50 (March 1985).
35. True, G. F., Glass reinforced cement permanent formwork, Concrete Society Current Practice Sheet No. 97, *Concrete*, 31–3 (Feb. 1985).
36. Rosenthal, I. and Bljuger, F., Bending behaviour of ferrocement-reinforced concrete composite, *J. Ferrocement*, 15(1), 15–24 (Jan. 1985).
37. Weyers, R. E. and Cady, P. D., Cathodic protection of concrete bridge decks, *ACI J.*, 618–22 (Nov.–Dec. 1984).
38. Marsden, A., Special reinforcing steels, Concrete Society Current Practice Sheet, *Concrete*, 19–20 (Sept. 1985).
39. ICI Linear Composites Ltd, Basic physical properties of Parafil ropes, Harrogate (April 1983).
40. Burgoyne, C. J. and Chambers, J. J., Prestressing with Parafil tendons, *Concrete*, 12–15 (Oct. 1985).

8

Prestressed Precast Concrete Bridges

H. P. J. TAYLOR

Dow-Mac Concrete Ltd, Stamford, UK

8.1 INTRODUCTION

To completely cover the topic defined in the title of this chapter in the space available would be impossible. It is not the author's intention to produce a manual for design, considering all aspects from the selection of the scheme, through analysis and design to construction. This chapter is intended to give a review of the subject and to discuss current developments in some detail. Finally, future trends are suggested with, as justification, the historical context. In the review, references are given to papers that were of significance in their time and to sources for detailed explanations of subjects which can only be touched upon in this short text.

For the purpose of clarity and organisation of the text it is necessary to make some definitions of terms which are often used loosely in the literature. The most important are those which relate to span:

Short span	Bridges with spans up to 20 m
Medium span	Bridges with spans between 20 and 35 m
Long span	Bridges with spans exceeding 35 m

These definitions are used consistently in this chapter as the spans are particularly relevant in precast bridges with respect to the type of construction and the convenience of transport.

Finally, this chapter deals essentially with precast bridge decks; substructures are not considered.

247

8.2   HISTORICAL REVIEW

This review presents a simple account of the development of concrete
bridges, concentrating on the period after 1945, when the major precast
development took place.

8.2.1 Market

The type of bridges built in any country relate to a number of factors, of
which some of the most important are:

> geography;
> social history;
> material availability;
> technical development;
> fashion.

In the UK, the history of bridging developed with its industrial
development. First there were bridges over rivers, where fording was not
possible. The canals had a requirement for aqueducts, some of which are
quite long, and for road overbridges of short span. The construction of the
railways brought big changes, the development in large numbers of cast
and wrought iron bridges as well as bridges of other materials, timber,
stone, brick etc. In this century, the dramatic expansions of motor transport
and the consolidation of the rail routes created a demand for many short
and medium span bridges using, at first, steel and then, after 1945, concrete
as the dominant material.

Britain does not have many wide rivers, and the number of long span
bridges was quite small until quite recently when road crossings of the
wider rivers and estuaries were built. Even now the demand for bridges with
spans over 100 m is likely to be small.

The bridge market is and has for the last hundred years been dominated
by a limited number of clients, firstly the canal and railway companies and,
in the UK, now by the Department of Transport (DTp), the Welsh, Scottish
and Northern Ireland Offices, the Local Authorities and British Rail (BR).
Both the DTp and BR have used their large resources to encourage
standardisation of approach, construction, analysis and design as much as
possible.

Standardisation is discussed later in this chapter, but it is sufficient here
to say that the story of precast concrete bridges is the story of
standardisation. The definition of design criteria has developed from the
normal code and standard creation process, with the possibility of open

debate between all interested parties. These basic criteria have then been used as a basis for client-defined criteria, particularly with respect to the DTp. Consultation with interested parties is still carried out.

The nature of the types of bridges required in the last 40 years, mainly bridges for road construction and improvement, and the method of design and procurement, have had a big influence on the development of prestressed precast bridges. In the last few years there have been changes which will make significant differences to the kinds of bridges being designed and built.

More bridges will be designed in the offices of consultants than before. This may make changes in two very different ways. On the one hand, the consultants, in order not to upset their major clients, may use the criteria rigorously and will not see a need for design development. On the other hand, because of their freedom, they may be an engine for change, experimentation and development of new ideas.

The economic balance between the construction materials and methods is now being questioned more than ever before, and a new breed of bridge designers, working with contractors in alternative design, are also starting to make their mark. Competition is likely to accelerate this trend.

In the future, therefore, we are likely to see much more competition in the bridge market than we have seen in the past 20 years. The precast concrete bridge has performed well so far and is going to continue to be a significant force.

8.2.2 Materials

When the technique of prestressing started between the two world wars, an important limiting factor to development was the availability of suitable concrete and steel. The concrete in particular was not of sufficient strength or stability to be capable of being used effectively with high strength steels. The steel could not be stressed sufficiently because of the low concrete strength and because of the high creep and shrinkage of the concrete; strain loss in the steel was considerable.

The growth of the strength of concretes is shown in Table 8.1. The strengths shown are the maximum quoted code values. In reality, the users of prestressed concrete developed higher strengths and these are shown in the table where records are available. These strengths are needed for early transfer of prestress and good mould utilisation. For design, grade 50–60 is the general optimum in composite construction.

Initially, concrete mixes were designed to have very low water/cement ratios and were compacted with a high degree of vibration. The standard

TABLE 8.1
Concrete Strengths (Code Maxima for Prestress)—28-day Standard Cube Strength

Date	Code	Cube strength, N/mm² (psi), works	Contractors' potential
1933	DSIR*	26 (3 750)	
1948	CP 114*	31 (4 500)	55
1957	CP 114*	31 (4 500)	
1959	CP 115	52 (7 500)	60
1962	MoT Memo 785 DoE Memo 1118	52 (7 500)	
1972	CP 110	60	
1984	BS 5400	60	75

* Reinforced concrete codes.

bridge sections were designed so that steel moulds could be used which could transmit this vibration, often applied externally. With the increased amount of secondary steel and cast-in fittings that are specified now, mixes are usually placeable only if a plasticiser is used to increase workability.

The development of prestressing steels in the last 30 years has seen the change from wire being the only material for pretensioning to round wire strand and later to deformed (Dyform or Compact) strand. Round wire strand was introduced in 1953 in 19-wire form. The more common material used in pretensioning, 7-wire strand, was introduced in 1956. Finally, in the 1960s, Dyform strand was introduced. By 1970 the use of strand exceeded that of wire, and wire now is almost totally restricted in use to precast floors and other small prestressed products. The modern prestressed bridge beam, in the UK, therefore uses 7-wire round wire strand or drawn strand, according to the designer's choice. Both types will be low relaxation Class 2 to BS 5896.

The choice of strand can relate to the number of strands required in an element and the need to distribute these in the section but, in general, longer span beams are more often designed with drawn strands as this produces less congestion in the mould and generally leads to a lower percentage of final loss of prestress. Figures 8.1, 8.2 and 8.3 give data on transmission length and relaxation provided by one manufacturer.

8.2.3 Analysis

Analysis is only covered briefly in this chapter. The history of the main methods is of interest, however, as it enables the reader to fully understand

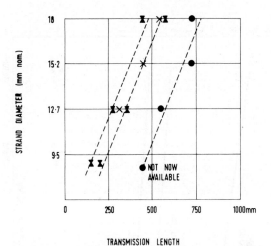

Fig. 8.1. *Transmission length of strands.* **X**, *C & C A tests, standard round wire strand (70% initial stress);* **X**, *standard round wire strand (75% initial stress);* ●, *Dyform strand (75% initial stress).*

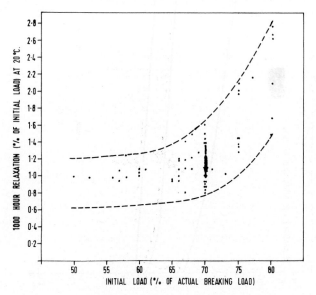

Fig. 8.2. *Relaxation of strand related to initial load.*

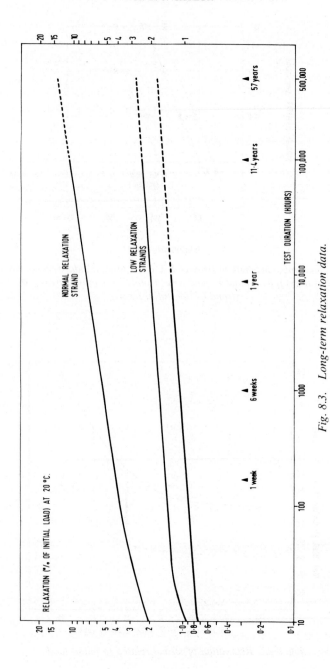

Fig. 8.3.   Long-term relaxation data.

the analytical context of the practical developments described in more detail. Hambly[1] has written a modern review and account of the various common techniques of bridge deck analysis, which is an excellent starting point for a study of the subject.

The growth of the use of precast bridges since the mid-1950s has been parallelled in the growth of methods of analysis. In the late 1950s simplified methods of analysis for isotropic and pseudo-isotropic plates were available. The load distribution theory[2-4] was developed in the next few years to the extent that it became the standard method of analysis for precast bridges until the use of the computer became widespread. Rowe[5] in 1962 published his book on concrete bridge design which covered load distribution theories and briefly dealt with other approaches, finite difference analyses for skew slabs and grillage analysis, which were to feature more prominently in the future. With the advent of the computer, the solution of many simultaneous linear equations became feasible and grillage and finite element methods became prominent.

Grillage analysis is well written up and is now the most popular method for slab-and-beam and slab bridges.[6,7] The grillage method has very wide application, can give useful results for a variety of structures and has a bank of good computer programs. Modern programs have good input and post-processing routines and present the chosen grid visually for checking. Grillage programs without these features are difficult to use and should be avoided.

Box beam bridges and flexible voided slabs are designed by a variety of methods, depending on the degree of cross-sectional distortion. Maisel[8] has produced a thorough review of the methods available.

8.2.4 Construction

Although some notable prestressed concrete bridges were constructed before 1945, in the UK the growth of prestressed precast concrete bridge construction took place later, with the establishment of pretensioning factories. Post-tensioned bridges were often cast on site, either next to the works or, in the case of Fig. 8.4, in a fixed central deck shutter. For this bridge, cast in 1950, the beams were cast on a pallet on trestles spanning the river and were then moved sideways to their final location in the deck. When all the beams were cast, the trestles were removed from the central shutter. The *in-situ* deck was constructed with reinforcement passing through transverse holes in the top flanges of the beams. Another interesting feature shown in the figure is the precast end block containing the Freyssinnet anchor cones.

Fig. 8.4.   Bridge construction, 1950.

Fig. 8.5.   Three-span box bridge constructed in 1948.

Figure 8.5 shows an early box beam bridge, constructed in 1948. The beam cross-section is shown in Fig. 8.6. The central void was formed with a collapsible timber shutter.

Manufacturers gradually developed their designs and moulds to the extent that the first advertised standard precast sections became available. Figure 8.7 shows such an early bridge, of 17 m span, constructed in 1957. This section is still available and is an extremely successful and economical design for infill short span decks.

In 1949 a group of interested engineers formed a committee named the Prestressed Concrete Development Group. This group, the PCDG, was instrumental in developing the first standard prestressed sections that were available competitively, from more than one manufacturer.

The sections, an inverted T, an I and a box section,[9] are all still available, although the I section is now seldom used and few manufacturers have standard moulds. The box section has also been developed and changed slightly to the current standard shape. These sections are shown in Fig. 8.8. An interesting inverted T bridge is shown in Fig. 8.9. This bridge,

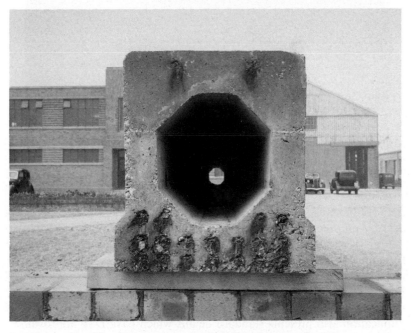

Fig. 8.6.   Box cross-section in bridge shown in Fig. 8.5.

constructed in 1969, has recently been the subject of a follow-up reassessment research programme which, together with the results of long-term loading tests on beams retained in the laboratory, confirmed the basic accuracy of the original design criteria.

In the mid-1960s the M beam was developed, jointly by the Cement and Concrete Association (C&CA) and the DTp. This beam, originally conceived as being used to produce a voided deck with an *in-situ* solid slab,

Fig. 8.7.   Early bridge with precast inverted T-type beams.

is one of the most common beams used to this day. The design of M beam decks is considered in more detail later in this chapter.

Finally, in 1973, the current U beam was introduced by Dow-Mac Concrete Ltd in conjunction with G. Maunsell & Partners. This beam is the latest addition to the standard bridge beam range. The Association of Bridge Beam Manufacturers has recently, with the backing of the DTp, produced a brochure with the full range of standard bridge beams.[10] This publication also presents standard details which have DTp approval.

The theme of standardisation was carried forward by the Department of

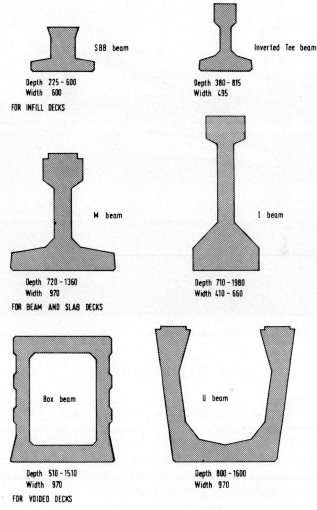

Fig. 8.8.   Precast bridge beam types.

Transport to the design of complete standard bridges. In conjunction with the DTp, a series of standard bridges were designed and the specifications and drawings made available for general use.[11] The bridges described by the DTp cover the common precast options of inverted T, M and U beam decks. The use of these standard bridge designs has been gradual, and some U beams are now being supplied by the manufacturers for these standards.

Fig. 8.9. Lightweight concrete inverted T bridge constructed in 1969.

8.3   DESIGN CRITERIA

8.3.1 Design

Some of the early design codes used in the UK for prestressed bridges are mentioned in the Introduction. Bridge design is now to the Bridge Code BS 5400 and as further defined by the client, DTp, BR etc.

The basic criteria in BS 5400 are little changed from those in the original prestressed concrete code, CP 115. BS 5400 is written in limit state terms with the following limit states being considered:

> collapse;
> serviceability;
> vibration.

The serviceability limit in prestressed concrete is defined by the need to eliminate or control cracking and the resulting penetration of the concrete by water and carbon dioxide. The exact processes of corrosion are unclear, but it is sufficient to say that a cracked structure will suffer from corrosion and degradation more rapidly than an uncracked one. For this reason, and because of the widespread use of salt for de-icing, major bridges are likely to continue to be designed to the Class 1 prestress limits for critical loadings. Some economies of steel without loss of durability would be possible if Class 2 were adopted, as Class 2 structures rarely, if ever, crack. This approach of allowing Class 2 design to the less critical load cases is now sensibly allowed in BS 5400.

Class 3 design is never likely to be adopted as, with current high steel strengths, a Class 3 structure will usually require more steel to satisfy the ultimate limit state than that needed for the serviceability limit state. As prestressing steel is less expensive at first cost than reinforcing steel, in terms of cost per unit of tension stress carried, and is easier to fix in a factory, the additional steel to satisfy the ultimate limit state will be prestressing steel. If more prestressing steel is provided than the minimum for serviceability reasons, it is logical to tension it and convert the Class 3 design to Class 2.

Prestress losses were a limiting factor in early prestressed concrete design. Recently, research has been carried out to measure the cracking strength of post-tensioned cast *in-situ* and pretensioned bridge beams of reasonable ages.[12] The results show that the beams after 20–30 years have good cambers, low levels of carbonation (1–3 mm) and no tendon corrosion. Loss of prestress is very similar to that which would be predicted by modern codes.

The exact computation of prestress loss is difficult and largely unnecessary. The loss of serviceability of a bridge with a large prestress loss error, say 50%, is likely to be very small and might change a Class 1 design to Class 2. The effect of inaccurate assessment of loss on the ultimate strength of a bridge is almost negligible. The effect on shear is likely to be greater than on flexure, but modern bridges now have many more nominal links than were used previously.

Despite this lack of criticality of loss, many designers go to excessive lengths to compute prestress, particularly in the various stages of construction of a continuous bridge made from prestressed single span elements. It is sufficient in these cases to make reasonable assumptions as to the age of the beams when they are incorporated in the structure, say one month, and the age when the continuity is fully effective, say two months. Such assumptions, with assumed creep and shrinkage relationships, Fig. 8.10, will be sufficient for most purposes.

8.3.2 Materials
Material durability and stability is a vital design consideration, since both concrete and steel reinforcement can degrade within the design life required

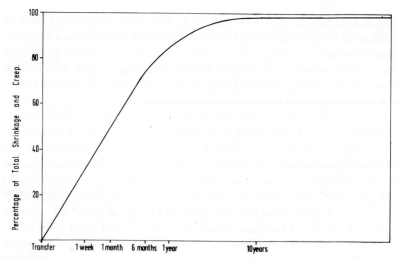

Fig. 8.10.   Typical rate of gain of creep and shrinkage loss in steam-cured bridge beams.

of bridges, 120 years for DTp bridges. Concrete durability has not been uniformly good. In *in-situ* construction, and particularly in substructures, the deterioration of concretes from alkali silica reaction (ASR), or from spalling from corrosion of reinforcement through carbonation and chloride contamination, is being increasingly reported. The precast pretensioned bridge beam has behaved in an excellent manner and, so far, no case of ASR has been reported, this despite the use of high cement contents in these units in the past. Neither have there been widespread problems with chloride-related corrosion.

The excellent performance of pretensioned prestressed beams is probably because of the low permeability of the concrete used, a consequence of the need for high early strength for quick transfer, proper curing and the need for good concrete mix design to cast the difficult narrow sections. Another important factor is the stability of the suppliers and the skill that can be built up in their long-serving workforce.

It is likely that in the future there will be demands to consider other criteria of durability than those used now—cement content, water/cement ratio and cover. This will produce difficulties with the development of tests, their reliability, the result and significance of non-compliance, remedial treatment and the contractual framework involved. These will be extremely difficult to resolve and, in the case of pretensioned precast beams, there is no proven need to improve or better define durability.

The use of calcium chloride or high alumina cement has not been common in bridge structures, and poor durability from these causes is not likely to occur.

The new requirements to limit total alkalis, or, if this cannot be done, to proscribe certain aggregates because of potential reactivity, is sensible and, with a mature view which will come as more research is concluded, problems from ASR will be controlled. No case of ASR in precast pretensioned beams has been reported.

Most prestressed concrete manufacturers use water-reducing plasticisers or superplasticisers in their mixes so that the mixes with the required low cement content (because of ASR) and low water/cement ratio can be properly placed in the thin and congested webs and flanges of bridge beams. The preceding approach will continue to ensure that precast beams are free from ASR. The use of sulphate-resisting cement is an unnecessary and harmful approach. The problems of mix design, particularly of workability and early strength, make such cements inappropriate for use in bridge beams.

The efficient and proper grouting of tendons in prestressed concrete work is essential for good durability. Instances have been reported of severe tendon corrosion in improperly grouted ducts, and some tendons have failed. Good grout design and practice are therefore essential. It should be remembered, however, that the problem only exists in post-tensioned work. Pretensioned bridge beams, where the concrete is cast around the tensioned strands, do not and cannot have any problem in this respect.

8.3.3 Specification
Bridges in the UK are most commonly designed to the workmanship clauses of BS 8110 and BS 5400 and to the client's specification. Both DTp and British Rail have comprehensive specifications and explanatory documents. It should be realised that much of the material in these specifications is intended for *in-situ* construction and may not be fully appropriate for precast construction.

8.3.4 Quality
This is an area where changes are being made. Bridges have traditionally been procured under the ICE Form of Contract, or variations of it prepared by the client. This has usually meant that there is inspection and checking of the design in-house or externally, and of the site and factory works by the resident engineer. In the case of factory production of standard units, this repeated checking is largely unnecessary, particularly of

items like mixer and instrument calibration etc. The trend towards quality assurance (QA) should make savings in this respect.

The DTp has already introduced a formal system of checking both the design type and detail design of bridges. Category 1 and 2 bridges are checked in-house and certified as having been checked. This is a form of quality assurance. The DTp is also encouraging beam manufacturers to start QA schemes. In construction work, QA is more completely applied. In this context, QA is a complete documented internal system, checked independently by an authorised third party organisation. The system covers all aspects of BS 5750 from the receipt of material and instructions to the testing and delivery of the finished product. Some manufacturers already have QA systems in operation and look forward to savings of time from the lack of need for client inspection on each job. At the same time, they anticipate the sales advantage of an assured product. So far, engineers for the major client, the DTp, do not seem to distinguish between the QA and externally supervised quality control systems, and still inspect everything. It is hoped that this continual waste of effort will be corrected in the future.

8.3.5 Transport
This is a design criterion which is rarely considered by an engineer outside the precast industry. It is, nevertheless, important: an unwise choice of element size or weight can result in unnecessarily high transport costs.

The regulations concerning the transport of unusual loads are complex and are subject to change by the authorities from time to time. The following, simplified, advice is based on the law at the time of writing. It is unlikely, however, that the rules will change significantly in the immediate future. In critical cases, the designer is advised to call a haulier or the Police for up-to-date advice.

The rules for transport of goods relate to both dimension and weight (length, width, height and weight). These apply to the vehicle and goods as one unit.

Basic loads

32 tonne truck	20 tonne payload
38 tonne truck	24 tonne payload

Unusual loads
Length
Under 38 tonne gross, up to 15·5 m—no restrictions.

From 15·5 to 27·4 m—notify the Police and other authorities and beyond 18 m have a driver and mate.

Over 27·4 m—find a route, obtain dispensation from DTp and have a driver and mate.

Width
Under 2·6 m—no restrictions.
From 2·6 to 3·5 m—notify the Police and other authorities.
From 3·5 to 4·1 m—notify the Police and other authorities and have a driver and mate.
From 4·1 to 6·1 m—obtain dispensation from DTp and have a driver and mate.
From 6·1 m—obtain dispensation from DTp, notify Police and other authorities and have a driver and mate.

Height
Under 4·2 m—no restrictions.
From 4·2 m—find a route and notify the Police and other authorities.

Weight
Under 32 or 38 tonne, depending on vehicle—no restrictions.
From 38 to 152 tonne—find a route, notify Police and other authorities and have a driver and mate.

Practical limits for the design of bridge beams which will minimise transport cost are:

   (i) Unit weight to be divisible into 20 or 24 tonne with as little remainder as possible.
   (ii) Length to be kept below 27·4 m.

Practical limits for the design of larger units, for example box bridge segments, having made allowance for the vehicle, are:

   (i) Width $< 3·5$ m.
   (ii) Length 18 m.
   (iii) Height 3·3 m.
   (iv) Weight 80 tonnes.

These rules are not proposed as hard and fast rules; in fact, bridge beams of lengths greater than 27·4 m are regularly transported. The rules will, however, minimise transport costs. These rules are illustrated in Fig. 8.11.

8.3.6 Service Life

The service performance of bridges is an important design criterion. The need for inspection, repair or replacement of out-of-life components, waterproofing, bearings etc. has made it necessary for the designer to

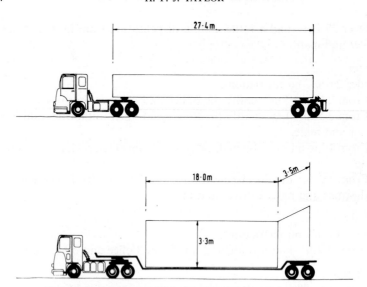

Fig. 8.11.   Practical rules for economical transportation.

consider the means of inspection and maintenance in his scheme. For this reason, access to critical areas must now be considered. This has resulted in halving joints away from supports of bridges becoming unpopular although, as discussed later, there are ways of overcoming this.

There is occasional discussion in the technical and non-technical press on the need to consider demolition in the design of new works. There is no proven case for the argument that the demolition of post-tensioned grouted or pretensioned beams is dangerous. In fact, the careless demolition of an arch bridge by the removal of fill from one side only is more likely to result in a buckling explosive failure than the use of normal concrete breakers on a prestressed beam. A recent research programme[13] has been carried out on this subject, and the results justify this view.

Post-tensioned ungrouted tendons can present difficulties in demolition as in corrosion protection during service life. Relatively few bridges of this type have been constructed, and it is likely that the use of the practice will become even less common.

8.4   BRIDGE TYPES

A large range of concrete bridge types has been developed. Many of these types involve prestressing, but the contribution of precasting is variable. In general, the bridge type is related to providing maximum economy of use of material and of construction technique for the particular span and application. Shorter spans have precast components as beams and, for transportation reasons, the longer spans use concrete slab elements or segments.

Figure 8.12 shows a range of types of bridge related to span. The bridge types in this figure are discussed in this chapter, but it is important to recognise that the figure does not completely categorise all structural possibilities.

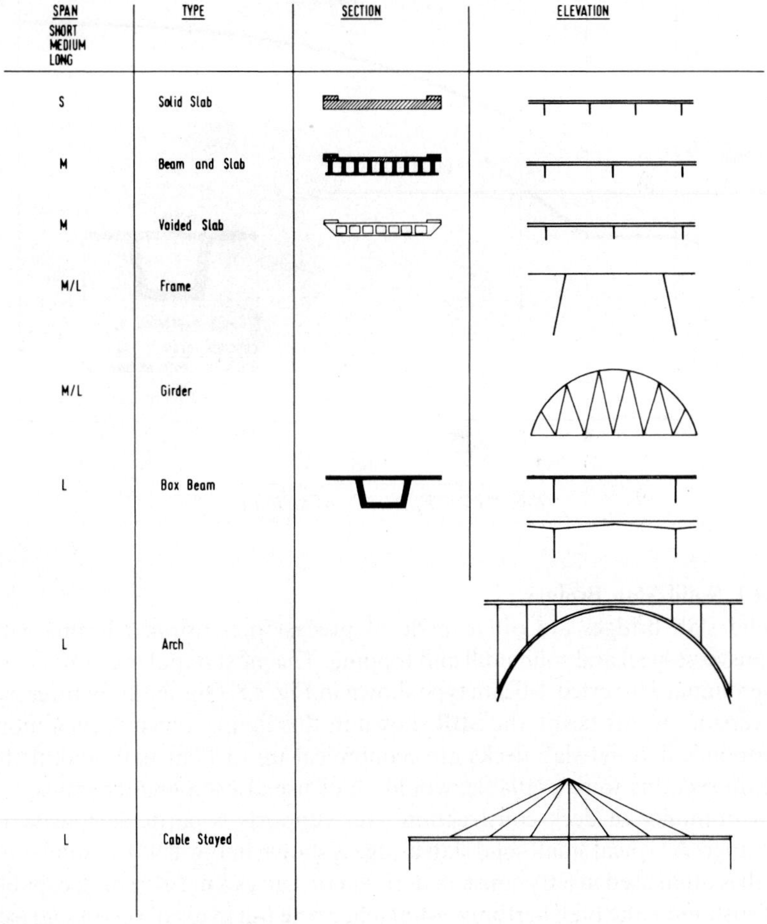

SPAN SHORT MEDIUM LONG	TYPE	SECTION	ELEVATION
S	Solid Slab		
M	Beam and Slab		
M	Voided Slab		
M/L	Frame		
M/L	Girder		
L	Box Beam		
L	Arch		
L	Cable Stayed		

Fig. 8.12. Bridge structure types.

One way of looking at the range in a logical, unifying way is to consider the relative efficiency in the use of materials in the various types. In a bridge, dead load is an increasingly important factor as span increases. Figure 8.13 gives an illustration of this and shows the percentage of the cross-section of a bridge deck that is voided, related to gross area at the support (deck width × depth).

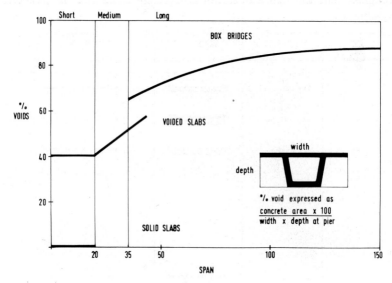

Fig. 8.13.   Efficiency of structure.

8.4.1 Solid Slab Bridges

Solid slab bridges are often made of precast pretensioned beams with transverse steel and solid infill and topping. The most popular beams are of the standard inverted T beam type shown in Fig. 8.8. One manufacturer has a version of this beam, the SBB shown in this figure, which is even more economical. Solid slab decks are economical up to 17 m span and, if the beam sections were available, would be of use at even greater spans.

Continuity of deck construction over supports is particularly easy to arrange. A typical small solid slab bridge is shown in Fig. 8.9. The solid infill slab is often used in jetty construction, where it gives a massive deck capable of sustaining the high berthing, point and crane rail loads often present (see Fig. 8.14).

Fig. 8.14.   SBB type beams in jetty construction.

8.4.2 Beam and Slab Bridges

A number of beam and slab systems are available, based mainly on M, I and U beams (see Fig. 8.8). The beam and slab bridge allows some economy of dead load and, with the practical limit of transport being 33 m in Europe and slightly longer in the USA, this type is the most common medium span bridge. The form of construction requires the bridge beams to be placed side by side, with a range of spacings, and a top deck cast on removable or permanent shuttering. A large T section is popular in the USA but is not used in the UK, except for car parks. Its unpopularity for bridge design in the UK is probably because of its economy in the longer span ranges, which are beyond transport limits, and in the relative lightness of American loading.

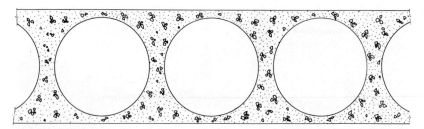

Fig. 8.15.   Voided slab with 60% voids.

8.4.3 Voided Slab Bridges
Voided slab decks tend to be cast *in-situ*, although the M beam was first envisaged as being used with an *in-situ* bottom flange. Box beams are also effectively voided slabs, so they are usually placed contiguously and are cast or stressed together laterally. Figure 8.15 shows the problem, with voided slabs, of sufficiently increasing the void percentage. The nature of a typical void former, and the difficulty of effectively casting underneath it, makes casting a slab with more than 60% voids very difficult.

8.4.4 Frame Bridges
Frame bridges are often used for grade separation in motorways. They can be reinforced or prestressed and are most often cast *in-situ*.

8.4.5 Girder Bridges
Concrete girder bridges these days are a rarity. Figure 8.16 shows a very attractive bridge of this type where prestressing was used off-site in component manufacture and on-site to assume full monolithic action. This bridge has a span of 64 m and was constructed in 1958.

8.4.6 Box Bridges
As well as its use in short and medium span slab bridges, the box girder provides precast concrete with its major opportunity. Box beam configurations and methods of construction are various and will be discussed later in the chapter.

8.4.7 Arch Bridges
For the right situations, where there is sufficient construction depth available, arches provide an economical solution for spans in the range 2 to 400 m. Arches are out of the scope of this chapter, but it is worth noting that

Fig. 8.16.   Prestressed, precast girder bridge, 1958.

there is an increasing market for the repair, strengthening or replacement of the many masonry arches used today.

8.4.8 Cable Stayed Bridges
A number of the shorter span cable stayed bridges are made in concrete. They are outside the scope of this chapter, however.

8.5   PRECAST CONCRETE IN BRIDGES

8.5.1 Infill Slabs
Infill slab bridges are the oldest of the design types described in this chapter. Precast concrete beams have been incorporated successfully in significant numbers into solid *in-situ* slabs for over 40 years. The design methodology was developed in the 1940s and 1950s to the extent that the method of approach described by Holland[14] survives to this day.

The typical deck layout of an infill deck is shown in the Bridge Beam Manufacturers' Publication.[10] The beams, of whatever type, are placed side by side with transverse reinforcing bars through web holes and are then infilled and covered with 75 mm or more of reinforced topping.

Load distribution methods are still used for analysis of these decks, but now grillage analysis is taking over, particularly when abnormal loading is specified. Originally, the following simplifications were used for analysis. Analysis for HA loading was, for skews less than 20 degrees, by consideration of a single strip. Transverse reinforcement was an arbitrary amount, justified by tests on model decks. For larger skews, tables based on finite difference analysis were used. HB loading was treated as an increase in standard HA loading for spans from 6 to 12 m, as described in the MoT Memorandum IM10.[15] This was only applicable where the deck continued to 1·5 m from the edge of the carriageway or where equivalent edge stiffening was provided. These methods are now, strictly, not allowed as the MoT documents are withdrawn, but are still very useful in preliminary design.

Grillage analysis has now replaced these simplified methods, and a number of very convenient programs are available. The choice of grillage layout is a matter of experience and judgement but, for these types of deck, a grillage member for every two longitudinal beams and transverse members with spacing of up to 1·5 times the longitudinal spacing will usually suffice. For complex skew layouts, it is wise to do one or two trial analyses with coarse grids first to get a feel of the problem.

The methods of design of prestressed composite beams are all well written up and the design criteria in BS 5400 are all clear. The designer may sometimes find one aspect of the design process rather tedious, requiring a number of trial and error checks. This is illustrated in Fig. 8.17. The code allows the infill concrete to carry some tension below the neutral axis. To actually determine the neutral axis position consistent with the value of f_t at the correct point with respect to strain compatibility is rather tedious. It is a sufficient and conservative simplification to assume that no *in-situ* concrete is carrying any tension beyond the first approximation of neutral axis, the mid-point of the slab.

The method of design for shear in BS 5400 is conservative but is not restrictive in practice. Recent research by Clarke and Evans[16] gives an

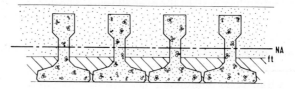

Fig. 8.17.   Solid slab deck—cross-section.

indication of the reserves of safety in the approach and, in the future, more refined methods of design should be developed.

The use of standard details as recommended by the Bridge Beam Manufacturers is important to produce a workmanlike design which can be realised as a structure without defects.

8.5.2 Beam and Slab Decks

Beam and slab decks with precast M, U and sometimes box beams cover the span range from 16 to 32 m. The design approach has been described by Somerville and Tiller[9] and by Manton and Wilson.[17] The use of the standard I beam is now very rare and is not discussed in this chapter.

M beam decks were originally envisaged as having a solid bottom flange with infill concrete and a top flange cast on lost formwork, as in Fig. 8.18(a). An end diaphragm was used in all cases. This method of construction was found to be expensive and, with the penalty of going to an extra beam depth, it was possible to eliminate the lower flange completely, greatly simplifying the construction (Fig. 8.18(b)). This is by far the most common application. With the new requirement for inspection in mind and a further desire for economy, some designers are placing M beams with gaps of up to 500 mm and more between them. This has the penalty of the need for a deeper beam and possibly a deeper slab, but does have the advantage that all the superstructure may be inspected (Fig. 8.18(c)). The cost benefit of having fewer beams, and the advantage of providing more area of the deck which may be inspected, may make this form of construction more popular in the future.

End diaphragms of M beam decks almost always have transverse steel through an end hole in the beam and neatly encase the ends of the beams. Typically, the top slab of an M beam bridge is 160–200 mm thick. The diaphragm stiffness will have an influence on the deck beams and will tend to develop hogging moments in some load combinations. The problem of the diaphragm providing too much stiffness in a skew deck, and developing too much reaction at the obtuse corners and too little at the acute corners, can be overcome to some extent by locally making the deck slab thinner. Another way of tackling this problem is to deliberately provide suitably detailed crack inducers in the diaphragm to reduce its stiffness. Research has been carried out on support diaphragms, and simple rules for the stiffness of these composite members are available.[18]

A number of methods of edge treatment may be devised, but unless a special edge beam or panel is used the problem of the unattractive elevation of the edge beam flange remains. Some manufacturers are able to supply a

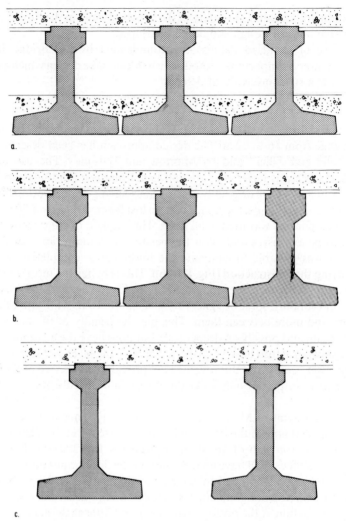

Fig. 8.18.   M beam decks: forms of construction.

special edge beam with an additional vertical web which produces an attractive side elevation (Fig. 8.19).

U beam decks have a lightness in appearance from the soffit which gives them great popularity; they also, with the inclined external web, have an attractive side elevation. The deck has an end diaphragm which need not extend to the full depth between the beams. This form of diaphragm is

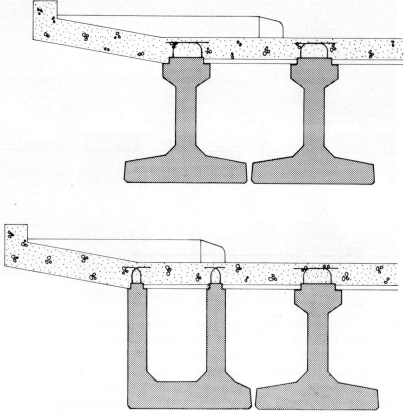

Fig. 8.19. 'UM' beam giving attractive edge detail.

particularly useful as it aids inspection of the bearings and provides a means for jacking up the bridge should replacement of the bearings be necessary. U beam spacings can be economical up to 2·5 m, and deck thicknesses are usually in the order of 160–200 mm.

U beams are manufactured with a mould with a collapsible core. For the longer spans it is possible, at some additional expense, to make a U section, for example U12, with a core from a smaller section, say a U10; this has an additional 200 mm of lower flange thickness which can accommodate extra prestress. Figure 8.20 shows a U beam deck during construction. The lost top shuttering and the support structure for a mobile edge form can be seen.

M and U beam bridges are usually analysed by grillage analysis. For M beam bridges there need not be a grillage member for every beam in the

Fig. 8.20.   U beam deck during construction.

bridge; West[6] gives guidance on this. A variety of approaches have been used with U beam decks. U beams may be modelled with one grillage member per beam or with one member per web.

Care must be taken in the assessment of the cross-members so that they truly relate both to the correct geometry of the deck and provide the correct stiffness. In some decks, carrying very high loading, the U beams are placed almost touching. In these cases, a 'shear key' model, similar to that used in a box beam deck, is appropriate.

8.5.3 Box Bridges

Box beam bridges have become dominant in the span range 35–250 m. Reference to Fig. 8.13 shows the efficiency of this form of construction with respect to structure weight.

Box beams may be cast *in-situ*—the shorter spans often are—but once spans become long and are consequently less accessible from below precast methods of construction become more common. Once a full shutter is dispensed with and a short segment of shutter is used to project or cast the bridge forwards precasting becomes an important option.

Figure 8.21 shows two common design approaches for box beam construction. The span by span method is appropriate for shorter spans

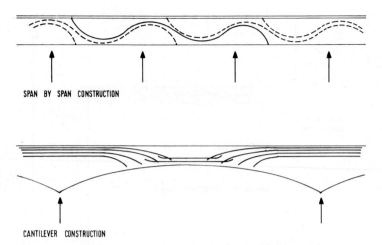

SPAN BY SPAN CONSTRUCTION

CANTILEVER CONSTRUCTION

Fig. 8.21. Approaches to box bridge construction.

and if many spans are involved. In this case, a span is cast and stressed in one operation back to the rest of the structure. This requires support for the unstressed *in-situ* or precast units over the whole span before stressing. Span by span construction has been carried out with precast segments, often with wide concrete-filled joints.

The cantilever construction method is used for longer spans and can also be of more use where fewer spans are required. This method of construction is often used with variable depth units and lends itself to the narrow joint counter-cast precast method of construction. Figure 8.22 shows the relationship between span and type of construction from a survey in the USA. The two methods, balanced cantilever and span by span, are shown in their commonly found and presumably economic ranges. The size of the shaded area gives the relative proportions of the construction types, and the length denotes the economic ranges. The figure also shows the common span range for an '*in-situ*' method of construction, incremental launching.

Box segments are cast for use with either the wide or the narrow jointing technique. With the wide joint technique, the units are cast without reference to each other and are then set up in their final location with a relatively wide joint, in excess of 75 mm. The joint is shuttered and made up with *in-situ* concrete to give the correct final bridge profile. Variations in joint width ensure good alignment. The narrow joint technique requires that the units are cast touching each other and are stressed together with a narrow glued joint. In this case, the alignment of the bridge is assured

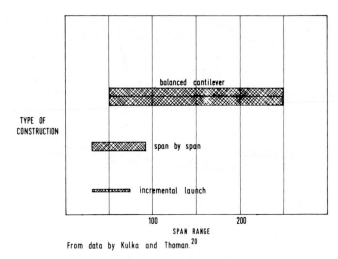

TYPE OF
CONSTRUCTION

balanced cantilever

span by span

incremental launch

100   200

SPAN RANGE

From data by Kulka and Thoman.[20]

Fig. 8.22.   Box girder bridges in the USA.

mainly by careful surveying and casting practice at the precasting site. Because of the number of units that are required in a scheme, careful detail design of the bridge is necessary so that a unit may be cast from the segment mould each day.

Conceptual design of box bridges has been researched, and useful feature surveys exist which give data on various section properties.[19,20] Detail design involves an understanding and consideration of the production techniques that are used. Figure 8.23 shows a section of a unit and part plan to illustrate some of these practical features. The unit itself is cast inside a steel mould with a collapsible steel core. Each unit is cast touching the previous unit, with careful checks being made to ensure correct alignment before and after casting. The prior unit is then taken into store and the later unit used as an end master for the next. In order to get a daily turnround of the mould, which is essential to construct the bridge in a reasonable timescale, the mould, and in particular the core, must be designed for rapid stripping, modification and setting up for the next cast. The figure shows typical positions for the 'blisters' needed to anchor the permanent tendons. The temporary stress needed to compress and control the joint is often provided by straight bars into the end face of the units, which are reclaimed when the permanent stress is applied.

The permanent stress blisters may not be provided in each unit; top and bottom blisters may be required in the units near to the centre of the span where the bottom chord continuity steel is needed. It is necessary therefore

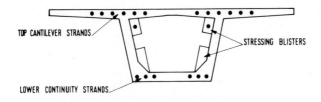

TOP CANTILEVER STRANDS

STRESSING BLISTERS

LOWER CONTINUITY STRANDS

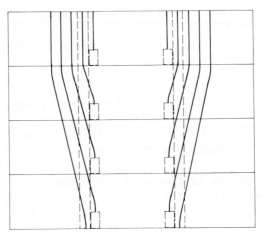

Fig. 8.23. Unit for segmental construction.

to have a mould which can rapidly be changed in this respect and in which there is a systematic means of getting prestress out into the flange as more prestress is built up (Fig. 8.23).

At the support of a box girder bridge it is common to provide thickening or deepening of the section at the point of maximum moment. Additional strength may be provided by deepening the bottom flange within the section or by deepening the section. Figure 8.24 shows a method of producing a deeper section at the support with the advantage of minimum mould modification. This method of deepening the section is, of course, much easier if the box webs are vertical. Webs may be thickened towards supports and this may be conveniently done in discreet steps of the core, internally.

A final difficult area to detail is the means of lifting the unit and supporting it from the overhead girder while the glueing and temporary stressing are carried out. Often the top flange is thin and lacks punching

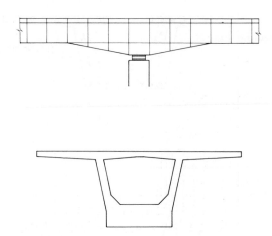

Fig. 8.24. Simple modification to unit at support.

strength for simple connection sockets. The top stressing blister provides a place where the flange is thicker, and may be used as the location of the lifting gear.

Analysis of box bridges, after the conceptual design stage and consideration of practical details, is simply carried out at first by simple continuous beam analysis. Detailed design can be followed up by grillage analysis and then by more complex methods. Maisel gives a thorough review of many of these methods.[8]

8.6 SPECIAL PROBLEMS

8.6.1 Half Joints

Half joints are not easily or economically formed in inverted T or M beams. The webs are too thin for sufficient reinforcement to be provided. The only way of solving this problem is to have solid, rectangular end blocks of the full soffit width. This requires that the outer mould sides, which are of steel and extend the full length of the prestressing bed, are cut to provide the end block. This is very expensive and ruins the mould side for future high quality work.

It is possible to provide half joints in box beams, and these beams are often cast in this way. They have the disadvantage that if they are put in a cellular deck there is no way in which the half joint area can be inspected. Bearings cannot be replaced and drainage channels cannot easily be cleaned. Box beams, spaced apart in a form of U beam deck (with vertical webs) do provide means for inspection of the bearings if they are on continuous supports, with shallow diaphragms between the beams. This is shown for a U beam deck in Fig. 8.25.

U beams have the same difficulty as M beams of being expensive to form half joints because of the need to cut the inner shutter. Recent research[21,22] has developed a means of casting a U beam half joint as a two-stage process with no mould modifications. The method is illustrated in Fig. 8.26. In stage 1, the beam is cast with the half joint formed in the webs, without cutting the internal, continuous, stressing line length shutter. Also, at this stage, reinforcement and sockets are provided to make a connection with the infill concrete of stage 2. In stage 2, after the internal shutter has been removed, the main half joint reinforcement is fixed and a series of ties are fitted which are designed to hold the webs together. It is these ties which wedge the infill concrete into the beam and ensure monolithic behaviour. The infill stage 2 half joint concrete is then poured, completing the half joint end. This form of half joint, with the shallow diaphragm between the beams, provides the solution shown in Fig. 8.25 and also satisfies the current criticisms of half joints with respect to access.

The method of half joint design given in BS 5400 proposes inclined reinforcement which carries the concentrated forces as a truss system most efficiently, as in Fig. 8.27(a). The alternative method, with vertical links and a horizontal tie at the half joint plane, is less structurally efficient (Fig.

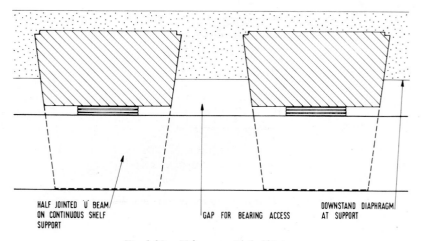

HALF JOINTED 'U' BEAM
ON CONTINUOUS SHELF
SUPPORT

GAP FOR BEARING ACCESS

DOWNSTAND DIAPHRAGM
AT SUPPORT

Fig. 8.25. U beams with half joints.

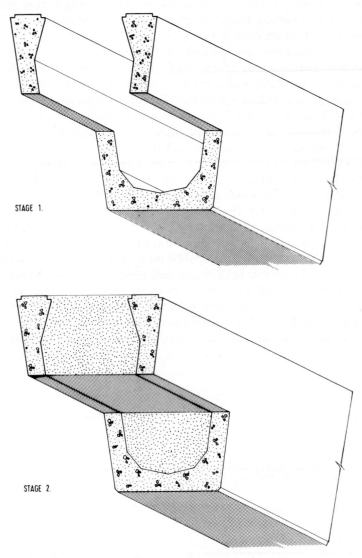

STAGE 1.

STAGE 2.

Fig. 8.26. Half joint casting process in a U beam.

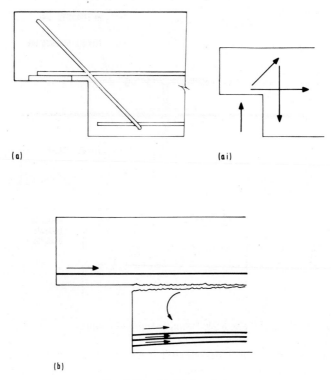

(a)

(ai)

(b)

Fig. 8.27. Half joint design.

8.27(ai)). It does, however, provide a second force method if the critical inclined bars are misplaced. It is recommended that an additional percentage of the vertical force is carried in this way, in the order of 25%. When the joint is in a pretensioned beam, the pretension, concentrated in the soffit, can produce horizontal cracks of the form shown in Fig. 8.27(b). The vertical links mentioned previously cross these cracks at the most favourable angle and add to the end block reinforcement needed to control the cracking.

8.6.2 Continuity

The difficulty of the design and maintenance of joints between spans of simply supported multi-span bridges has led designers to consider the use of continuity over supports more often. Continuous span construction is possible with all forms of precast concrete bridge construction, and practical details to solve the problem neatly have been developed.

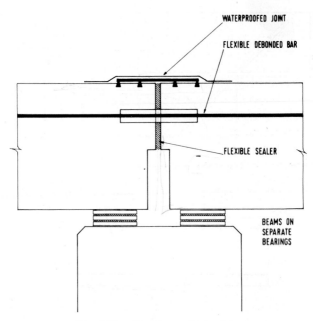

Fig. 8.28. Simple continuity joint.

One approach is to provide a flexible tie or strut between the spans which, whilst not carrying moments, controls the movement of the spans in such a way that a simple joint waterproofing system can be used. This is illustrated in Fig. 8.28. The bar may be a suitably protected reinforcing bar, debonded and soft wrapped over a sufficient length for the number of bars needed to provide longitudinal restraint to the deck whilst not having sufficient stiffness to carry moment and develop fatigue problems. The narrow joint is constrained in such a way that a straightforward water seal and waterproofing system will have a long life.

The method of approach requires that both decks are on bearings and therefore requires a rather wide column head. The method may be conveniently used on solid slab decks where the rubber bearing pads are not expensive or wide.

A design requiring full continuity over a support does enable only a single line of bearings to be used. The cross-head or support section may be long or short. Wide cross-heads have the advantage of extending the spans of the beam sections, although they produce heavy decks.

The use of an *in-situ* cross-head in negative bending produces some technical problems which are not satisfactorily resolved at present. In order to connect the deck together longitudinally in the construction phase and also to carry any sagging moments from creep and shrinkage, it is necessary to have some form of tie at the level of the bottom flange (Fig. 8.29). This tie may be in the form of a weld, a proprietary coupling, intersecting U bars or reinforcement between the bottom flanges of the precast units. The magnitude of this steel in the construction phase may be easily reduced by specifying shuttering with flexibility in the longitudinal direction or the use of good quality bearings for the temporary supports. In any event, it is wasteful to cut the strands at the beam faces as, if these are left long, they may be cast into the diaphragm, giving a useful extra tie. The strands may be specified to be of sufficient length to bond, but this may result in loss of economy by the manufacturer as the extra distance between the beams in the casting line may reduce the number in each cast. The manufacturers should be consulted about this at the design stage.

No matter what method of connection is chosen, it should be drawn out at a large scale to ensure that the site operations can be carried out. Swaging in particular requires considerable clearance around the bar.

The design of the cross-head in hogging bending presents fewer practical difficulties. The main problem is to decide upon the design approach for the precast beams. BS 5400 clearly states that the prestressed units should be considered to be prestressed in negative as well as positive bending: CP 110 said something similar and the handbook to CP 110[23] suggests that the section in hogging tension should be designed to Class 3. There appears to be no logical reason why a prestressed beam cannot be designed to carry reversed moments as a reinforced concrete section at a support if due allowance is made for the compressive prestress at the soffit in the cracked structure. At present this is not allowed in our design codes. More research is needed on this topic. Three design approaches to the problem are possible and are shown in Fig. 8.29. The compressive forces in the beams may be reduced by debonding, and the cross-head may be made of sufficient width to give no or negligible tension in the prestressed beam (i). The beam may be cast with an inclined or cut-down top flange to remove the material that would otherwise be in tension (ii). This approach gives more space for transverse steel in the cross-head, and is to be preferred. The beam with inclined webs shown in Fig. 8.7 is particularly useful in this respect. The fact that this approach is acceptable points to the fact that the whole exercise of eliminating tension in the missing area of the beam is probably unnecessary. Finally, the precast beam may be made with deflected tendons (iii). This approach conveniently provides compression at the top of the beam at its end. Not all the bridge beam manufacturers provide beams with

CONSTRUCTION METHOD

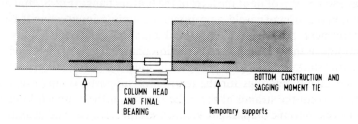

COLUMN HEAD
AND FINAL
BEARING

Temporary supports

BOTTOM CONSTRUCTION AND
SAGGING MOMENT TIE

CONTROL OF DECK BEAM TOP FIBRE TENSION

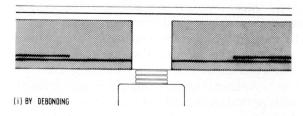

(i) BY DEBONDING

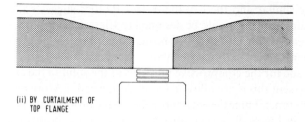

(ii) BY CURTAILMENT OF
TOP FLANGE

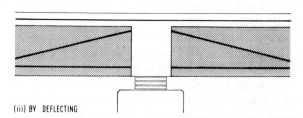

(iii) BY DEFLECTING

Fig. 8.29. Design for continuity.

deflected tendons, and the use of this approach may reduce the number of options for supply.

The design of the cross-head itself should now be considered. Two problems remain, the design for transverse bending if it exists and the design of the shear connection or support of the precast beams. Transverse bending may be carried by reinforcement or prestress and will depend on the extent of the support provided.

It is common now for these decks to be carried on two or more columns as point supports, increasing the need for proper analysis and detail design of the cross-head in the transverse direction. There is no need to provide transverse prestress to the cross-head area where it is connected to the longitudinal beams. Tests have been carried out[24] and have shown that the degree of composite action between precast beams and cross-head is excellent, without transverse prestress.

The final design problem is that of the shear connection between the *in-situ* diaphragm cross-head and the precast beams. Figure 8.30 shows the

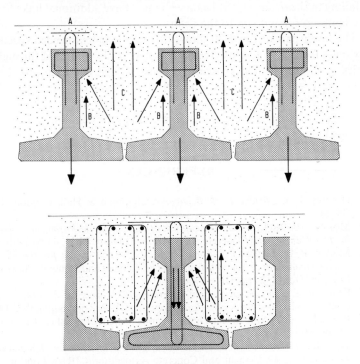

Fig. 8.30.   Design basis of support section with continuity.

cross-section of a deck at this point, with M beams incorporated into the diaphragm. Three methods of shear transfer are possible:

(A) The beams hang up on protruding links, anchored into the top slab and diaphragm concrete.
(B) The beams are supported by vertical shears to the composite web infill.
(C) The support for the beams comes from the concrete diaphragm which cantilevers out between them. This support can be designed using a truss analogy.

It is common to rely on method (C) to carry the shear forces; mechanisms (B) and (A) then provide extra security in their intrinsic load-bearing capacity. For method (C) to be effective, the negative bending top steel must be enclosed with links, which are anchored just above the bottom flange of the beams. The inclined thrust from this support then anchors beneath the top flange of the precast beams. The flange must then have links to prevent it failing in shear, and the M beam web must have additional links to carry the hang-up force down to the soffit.

The lower part of Fig. 8.30 shows the reinforcement needed specifically for these force interactions. Transverse steel is also required through end holes in the beams and around the back of their end faces, to stop the inclined forces from causing the beams to spread laterally. This is probably an extremely conservative method of approach, but it does have the advantage of simplicity of concept and enables the detailing to be thought out logically.

REFERENCES

1. Hambly, E. C., *Bridge Deck Behaviour*, Chapman & Hall, London, 272 pp. (1976).
2. Morice, P. B. and Little, G., *The analysis of right bridge decks subject to abnormal loading*, Cement and Concrete Association, London, Db11, 43 pp. (July 1956).
3. Morice, P. B., Little, G. and Rowe, R. E., *Design curves for the effects of concentrated loads on concrete bridge decks*, Cement and Concrete Association, London, Db11a, 24 pp. (July 1956).
4. Rowe, R. E., *The design of right concrete slab bridges for abnormal loading*, Cement and Concrete Association, London, Db12, 8 pp. (September 1958).
5. Rowe, R. E., *Concrete Bridge Design*, Applied Science Publishers Ltd, London, 336 pp. (1962).
6. West, R., *Recommendations on the use of grillage analysis for slab and pseudo slab bridge decks*, Cement and Concrete Association/CIRIA, London, 24 pp. (1973).

7. Cope, R. J. and Clark, L. A., *Concrete Slabs: Analysis and Design*, Elsevier Applied Science, London (1984).
8. Maisel, B. I. and Roll, F., *Methods of analysis and design of concrete box beams with side cantilevers*, 176 pp. (November 1974).
9. Somerville, G. and Tiller, R. M., *Standard bridge beams for spans from 7 m to 36 m*, Cement and Concrete Association, London, 32 pp. (1970).
10. Prestressed Concrete Association, *Prestressed concrete bridge beams*, 2nd edn, British Precast Concrete Federation, Leicester, 10 pp. (1984).
11. Department of Transport, *Standard bridges*, HMSO, London, 19 pp. (1979).
12. Buckner, S. H., Lindsell, D. A. and Robinson, S., *Prestress losses in full scale structures*, CIRIA Report, London (1986).
13. Lindsell, D. A. and Buckner, S. H., *Prestressed concrete beams: controlled demolition and prestress loss assessment*, Construction Industry Research and Information Association, London, 49 pp. (1987).
14. Holland, A. D., Prestressed units for short span highway bridges, *Proc. ICE*, Pt II, **4**(2), 224–97 (June 1955).
15. Ministry of Transport, *Standard highway loading*, London, Interim Memorandum—Bridges IM10 (1970).
16. Clarke, J. L. and Evans, D. J., *Vertical shear strength of composite beams*, Cement and Concrete Association, London, 25 pp. (March 1983).
17. Manton, B. H. and Wilson, C. B., *MoT/C&CA standard bridge beams*, Cement and Concrete Association, London, 20 pp. (1971, revised 1975).
18. Clark, L. A. and West, R., *The torsional stiffness of support diaphragms in beam and short bridges*, Cement and Concrete Association, London, TRA 510, 20 pp. (1975).
19. Swann, R. A., *A feature study of concrete box spine beam bridges*, Cement and Concrete Association, London (June 1972).
20. Kulka, F. and Thoman, S. J., Feasibility study of standard sections for segmental prestressed concrete box girder bridges, *PCI J.*, Chicago, Illinois (Sept./Oct. 1983).
21. Gill, B. S. and Clark, L. A., Collapse of concrete U beam half joints; theory and model studies, *Proc. ICE*, London, Part 2, **79**, 681–96 (Dec. 1985).
22. Gill, B. S., Clark, L. A. and Taylor, H. P. J., Full sized tests on concrete U beam half joints, *The Structural Engineer*, Part B (June 1986).
23. Bate, S. C. C. *et al.*, *Handbook on the Unified Code for structural concrete (CP 110: 1972)*, Cement and Concrete Association, London (1972).
24. Sturrock, R. D., *Tests on model bridge beams in precast to insitu concrete construction*, Cement and Concrete Association, London, 17 pp. (Jan. 1974).

9

Joints and Substructures

H. Aizlewood*

*Formerly Chief Bridge Engineer, West Midlands County Council,
Birmingham, UK*

9.1 INTRODUCTION

At first sight, it might appear that joints and substructures make strange bedfellows. Perusal of the pages which follow will demonstrate, however, that there is a close association between the use and performance of joints and the performance of substructures, arising not least from our inability to guarantee the sealing of movement joints. It follows therefore that the fewer joints we have, the fewer causes there are for deterioration of the structures.

Since the early 1970s, the increasingly widespread use of computer analyses of bridge decks has enabled the effects of differential settlement of the supports to be evaluated more reliably. Previously, very long and arduous manual assessments had to be made. These were usually prohibitively costly in cash terms, but more particularly so in the design time taken when answers were needed quickly. This factor, together with UK national policies of cheap first cost, led to a general view that only where conditions of solid ground support could be guaranteed was it reasonably practicable to design multi-span structures in fully continuous form.

During the same period, standardised factory-made precast prestressed beams became comparatively cheap for shorter spans, eliminating the need for formwork and expensive supporting falsework.

These two factors together yielded a popular policy among bridge designers to 'keep everything simple', with simply supported decks and

* Present address: 158 Baginton Road, Coventry CV3 6FT, UK.

hence span-end joints. The problems which have been widely associated with failed movement joints have led more recently to a much more careful consideration of designing continuous multi-span decks with allowance for what would previously have been considered prohibitively large differential settlements of supports.

9.2 PROVISION FOR ARTICULATION

9.2.1 Movement Gaps

It is imperative in the design and construction of any bridge which has provision for articulation that the designer conveys through the detailing to the contractor and the resident engineer the size of the movements anticipated, and precisely how these movements are to be accommodated. In preparing these details, consideration must be given to civil engineering tolerances, particularly in the fixing of reinforcement in adjacent members and the fixing of shutters. Once these considerations have been taken into account, and the extent to which temperature is significant, the size of the gaps must be detailed clearly. Whether they are to be left open, or closed with permanent shutters or seals, should also be clearly stated.

There has in the past been much slackness in this area of bridge detailing and of site construction and supervision. It has been common during inspections to discover concrete cast hard against earlier concrete with no provision at all for relative movement. Elsewhere, joints have been given inadequate gaps, and rendered fascias have been carried continuously across movement joints. Cracking and spalling are the inevitable results which, although not usually structurally serious, are nevertheless unsightly and the cause of unnecessary maintenance expenditure (Fig. 9.1).

9.2.2 Movement Joints

Experience has demonstrated that, sooner or later, all movement joints will leak surface water onto the supporting structure in general, and the bearing shelf in particular. This leaking water is contaminated in winter months by dissolved road de-icing salt. As the wetting and drying processes alternate, so the concentration of salt on the bearing shelf increases, and the concrete is penetrated by moisture containing free chloride ions. The penetration is frequently made easier because the concrete has been hand-levelled, over-wet and over-worked, with laitance floated to a smooth upper surface upon which the bearings are to be bedded. Penetrations of concrete pier tops and side faces to the extent of 100–150 mm have been commonly recorded in the

Fig. 9.1. Inadequate clearance at movement joint. Damage to pier crosshead and fascia panels caused by lack of clearance in the movement joint.

UK after a mere 15 years of service, with consequent significant rusting of the reinforcement.

A movement joint has to perform two distinct functions:

(a)   to provide a stable road surface upon which traffic can travel safely; and

(b)   to provide continuity of the waterproofing membrane across the structural gap, which is of variable width.

The problem is that these two functions have a high level of incompatibility. In the early experience of joint maintenance and repair, which was frequent, efforts were concentrated on maintaining the stability of the road surface. This aspect is directly related to road safety and traveller comfort and is the one which the travelling public perceives. Waterproofing was treated as a secondary matter. After all, the concrete supports were designed to be permanent and effectively maintenance-free! Experience has shown otherwise. The dramatic degradation of reinforced concrete in substructures caused by chlorides and frost is at least of equal long-term importance to road surface stability, although perhaps not demanding of such immediate maintenance action.

It is perhaps worth noting here that the road surface stability aspect of bridge joints is strongly related to the severity of the traffic loading across the joint. On bridges carrying mainly light vehicles, joints perform well if carefully made. On busy roads the incidence of joint breakdown increases dramatically as the number of heavy axles increases, particularly in the wheel tracks of the nearside lanes.[1]

9.2.3 Buried Joints

A very high proportion of bridge joints have been of the buried type. In this context, a buried joint is one in which the movement gap in the structure is covered at waterproof membrane level by a flexible flashing which is bonded to the deck and buried below continuous road surfacing asphalt (Fig. 9.2). They are suitable for small movements and rely upon the ability

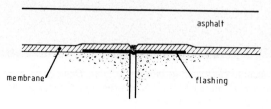

Fig. 9.2.   Buried joint.

of the asphalt (at least in its early life) to be flexible enough to adjust under the kneading action of passing traffic to the small movements anticipated. It should be noted here that the accommodation of structural movements within the asphalt assumes that these movements are slow, arising from creep and thermal movements. Significant dynamic movements are much too rapid to be absorbed. Buried joints therefore are effective only for short-span concrete bridges which are stiff dynamically—i.e. have very small live load deflections. However, in such situations buried joints can give good service within the working life of the road asphalt (Fig. 9.3).

Careful consideration of past failures leads to the following guidelines which are suggested for the many aspects of the installation and use of buried joints:

(a) Buried joints are suitable only for short-span, stiff concrete bridges, or for fixed ends, having maximum design movements of 7–8 mm without a steel plate, extending to 12–14 mm with a steel plate. The asphalt cover above the joint should be not less than 120 mm.

Fig. 9.3. Successful buried joint. The position of the joint is marked by the sealant in the kerbs and the crack in the thermoplastic white line. After 21 years of medium volumes of mainly light traffic, the asphalt remains uncracked, although there is some leakage of water. This is a viaduct fixed end joint, having dynamic rotational movements only.

Buried joints are unsuitable for skew angles greater than about 20°, or where the dynamic movements exceed 0·15 mm at the road surface.[1]

(b)   As with all specialist activities of installing bridge components, work should be carried out by an experienced sub-contractor. If the general contractor insists on using general site labour, the joint materials manufacturer or supplier must be on site during the installation work to agree standards of preparation and workmanship.

(c)   Where repairs are being undertaken (Fig. 9.4), the existing asphalt should be sawn parallel to the joint gap sufficiently far back not only to allow the reinstatement asphalt to be fully rolled, but also to remove all of the cracked and loosened asphalt around the damaged joint. Sawing is essential rather than using breakers with asphalt spades because these damage irreparably the standing asphalt faces. All materials between the saw cuts can then be removed to enable the substrate to be prepared.

(d)   Great care must be taken in preparing the substrate. Concrete surfaces on both sides of the structural gap must be sound, clean, dry, smooth and flat. These features are to ensure good adhesion of the flashing to the concrete, which is essential for waterproofing. Arrises must be straight and true to minimise the gap to be crossed by the flashing, and to avoid spaces for stones to press through. In the case of repairs to old joints, after removal of the debris, if the substrate is found to be not smooth and flat, it should be wire-brushed or sand-blasted to get it clean and sound. Very small irregularities and arris chips can be made good with polyester resin.

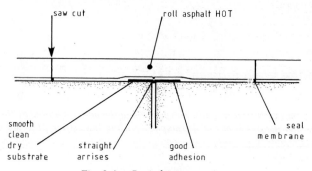

Fig. 9.4.   Buried joint repairs.

Larger depressions should be levelled with epoxy mortar or polymer concrete to achieve the smooth level standards required for new works.

(e) The substrate must be generously coated with the adhesive material and the flashing thoroughly bedded into it while it is still fresh. The flashing material will usually arrive on site in a coil which needs to be unwound carefully and flattened to eliminate wrinkles and bubbles when the strip is pressed into the adhesive. A small hand roller or squeegee will help with bedding the flashing strip. A shallow tuck or fold of flashing material pressed into the gap provides slack for when the gap opens. A seal between the flashing and the deck membrane must be made with care.

(f) If steel plates are used to bridge the joint gap, they should be 5–6 mm thick and in short lengths not exceeding 1 m. Longer lengths are difficult to bed and tend to rock in service. Plates should be galvanised, have downstanding lugs to locate them in the joint and all upper edges chamfered to avoid cutting the membrane materials. Plates should be thoroughly well bedded into mastic and covered with asphaltic membrane materials, the top layer of which should be toughened to withstand site traffic.

(g) Asphalt over the joint must be laid hot, preferably in good ambient conditions. Particularly with repairs, the quantities will be relatively small and will chill quickly. It should therefore be placed and rolled immediately it arrives on the site. Asphalt which has stood in a delivery truck, even double skinned, for an hour or so is unlikely to be hot enough to be laid successfully. Chilled asphalt cannot be rolled to optimum compaction, which is a fundamental requirement if premature breakdown over the joint is to be avoided.

(h) There is little evidence to indicate that a sawn crack inducer in the asphalt surface over the structural gap is of any use. To be effective, it would have to be at least 20 mm deep, which would weaken the asphalt structure and would be difficult to fill with sealant. It is probably better not attempted.

Attempts have been made to increase the tensile strength of the asphalt across the joint by introducing metal mesh or geotextile fabrics between basecourse and wearing course. These have suffered early failures because the metal is too stiff to remain bedded in the asphalt and the fabrics provide planes of lamination. Recently, the availability of polypropylene grids which can be used under or over the basecourse to strengthen it[2,3] provides a potentially possible method of improving asphalt performance over joints. A research programme at the University of Nottingham is currently in hand to study these techniques.

With all of the limitations and hazards associated with the construction of successful buried joints, the question will understandably be asked 'why bother?'. The answer is that the buried joint is comparatively cheap to install. Much more importantly, it is the only available joint system which allows the sub-surface, super-membrane surface water which flows over all surfaced bridge decks to flow uninterruptedly across the joint in a longitudinal direction. Every other available system creates a dam, which causes water to pond in the asphalt on the upstream side (Fig. 9.5). Under the action of passing wheel loads, hydraulic pressures are set up which are disruptive to the asphalt adjacent to the joint. The longitudinal drainage of other joint systems can be achieved in the short term, but no reliably long-term system is available. It should also be noted that the buried joint system needs no special preparation for resurfacing. This is not true of any other available system.

Fig. 9.5.   Dam caused by bridging joint. Subsurface water, mostly under central reservation, emerging on surface adjacent to rubberised bitumen joint.

9.2.4 Bridging Joints

During the late 1970s a different approach to small movement joints (up to say 20 mm) was developed and became available in the UK. It is here referred to as the bridging joint system. The design philosophy is based upon the replacement of the whole depth of road surfacing asphalt, together with the waterproofing membrane, for a short distance on both sides of the structural gap with an alternative material (Fig. 9.6). The replacement material has to be stiff enough to support the traffic, flexible enough to accommodate the joint movements, effectively waterproof, and must have good adhesion properties to the concrete on both sides of the gap. Various materials have been used including rubberised bitumen, polymer asphalt, polyurethane and acrylic resin. The first two of these have generally been not only the more successful but also both comparatively cheap.

Several UK companies market and install their own variations of the rubberised bitumen-based joint. The variations are principally in the make-up of the stone filler content and the way in which the joints are laid. The rubberised bitumen material is very soft compared with all the other systems. Joint width is therefore limited to about half a metre. Some deflection takes place under wheel loads, but recovery is quick. However, where a very high concentration of rapidly repeated heavy wheel loads occurs, some permanent deflection of the surface in the form of 'tracking' may occur.

Because joint movements are taken up by the internal deformation of the material, resistance to which is equal in all directions, the comparatively soft bituminous joints are suitable for skews of up to at least 45° (Fig. 9.7), and for joints with greater than normal vertical movements.

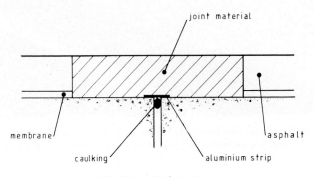

Fig. 9.6. Bridging joint.

Fig. 9.7.   Bridging joint with severe skew. This deck end has a stepped plan alignment. The original joint failed and was replaced with rubberised bitumen, together with patch repairs to the adjacent asphalt.

In the case of polymer asphalt, only one system is available in the UK. The increased stiffness of the material in service gives better support to road traffic, but some sacrifice of movement accommodation is the price. This is demonstrated by a tendency to form a crack between the joint material and the adjacent asphalt. This increased rigidity can, however, be used to advantage, particularly where an excessively wide joint is required. Examples are the repair of badly cracked asphalt around a failed earlier joint, and the covering of two movement gaps close together. There is effectively no upper limit to the width of the polymer asphalt used for a joint.

Because all of the bridging joint materials are effectively waterproof, a good joint seal is achieved as a part of the installation. The joint should not, however, be limited to the carriageway width only, but should extend to the full bridge width between parapets or as much of it as is practical (Fig. 9.8).

Every bridging joint system is marketed and installed only by the company responsible for its design and development. This is exactly as it should be for both new and repair works, and is not the case with any other joint system. Contractual responsibility for success or failure is thus placed clearly in the right place.

9.2.5 Nosings

For movements in excess of 20 mm, it is essential that the structural gap is continued through the road surfacing to appear as a joint on the surface. The choice of system lies between construction *in-situ* (nosings) or using prefabricated ('mechanical') joint units. The former will extend the movement range up to about 40 mm with a compression seal, or to 65 mm without a seal.[4] For larger movements, the use of prefabricated joints is essential. For very large movements, consideration should be given to a custom-designed and -built joint, particularly if there is a significant skew.

9.2.5.1 Nosings—General Considerations

Early *in-situ* nosings were formed of rigid materials such as concrete and rigid epoxy mortars. Provided that they are cast in short lengths of up to, say, 1·5 m and are securely bonded to the substrate, rigid nosings perform well. Indeed, a system using concrete containing chopped steel wire fibres is well known and used sucessfully in the UK (Fig. 9.9). The very many early failures were blamed in large measure on to the rigidity of these materials. A second generation of 'flexibilised' epoxy and polyurethane mortars was developed for bridge joint nosings. Because these materials can relieve themselves of their internal exothermal stresses shortly after casting, and

Fig. 9.8.   Bridging joint across central reservation. This rubberised bitumen joint was poured between shutters across the central reservation, and allowed to flow beneath the kerbs, to achieve a joint seal.

Fig. 9.9. Rigid nosing system. These nosings are of high alumina cement (HAC) concrete with chopped steel wire reinforcement in addition to mild steel anchorage bars. There is some breakdown of the polyurethane sealant bond in the main wheel tracks. Most joints on this viaduct are sound after nearly seven years of exceptionally heavy traffic.

the subsequent differential stresses developed during in-service temperature changes, they can be cast in continuous lengths to suit the site conditions. Whilst these materials have been developed largely for this purpose, it is questionable whether they should be considered to be truly civil engineering materials. They are very highly weather sensitive and demanding of the highest standards of conscientious workmanship. Placing in wet conditions or with a temperature less than about 10°C is not to be recommended. In addition, substrate preparation, batching, mixing, priming, placing, compacting, levelling, curing and sealing all have to be exactly right to ensure success. Whilst all of these operations apply also to concrete, there is no comparison between the sensitivity to variations of the two materials. It is hardly surprising, therefore, that even with these improved materials there have been many failures.

It has been observed from nosings in service that there are basically three modes of failure (Fig. 9.10):

(a) debond of the nosing block from its substrate, caused by inadequate preparation and/or priming, or by leaving too long between priming and placing;

(b) delamination of an upper layer of nosing from a lower layer, caused by too long an interval elapsing between placing and compacting the two layers;

(c) collapse of the nosing material, caused by incorrect batching or mixing of the constituents, by inadequate compaction of the mix within the trench, or by leaving a debonded nosing unrepaired so that it becomes broken up by traffic.

Unless a failure is caused by total debond, it is unlikely that failure along the whole length of both sides of a nosing-type joint will occur at one time. More generally, nosings are subject to partial failures while the remainder

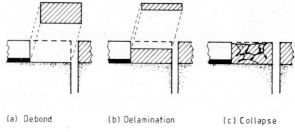

(a) Debond   (b) Delamination   (c) Collapse

Fig. 9.10.   Modes of nosing failure.

of the joint stays intact. In such cases, the decision must then be taken whether to repair and reinstate the damaged portion of the nosing, or to remove the undamaged part of the joint and replace the whole joint with a different material or system. The latter carries with it the very serious risk of significant damage to the deck end, and the consequent need to repair the reinforced concrete before the installation of an alternative joint system can proceed.

9.2.5.2 Sealants

All of the foregoing refers only to the matter of road surface stability. The question of joint seals is also of importance. Unless steps are taken to seal the structural gap at the deck end, there is little chance of effecting a durable seal with our present level of sealant technology. Rigid nosings in short lengths have planes of water passage transverse to their lengths. Compression seals are not effective in keeping out surface water, and they sometimes come adrift from the nosings with risk to passing traffic.

Poured mastic seals perform better than compression seals, at least in the short term, although their use may be limited to joints with movements less than 12 mm.[1] Polysulphides tend to age fairly rapidly. Polyurethane mastics perform better, but their bond to the nosings is suspect. This is not helped by the spalling of the arrises which is a feature of concrete nosings. There is some doubt also about the compatibility of pitch-extended epoxy nosings with polyurethane sealant. It is therefore desirable, if nosings are used, to effect a double seal, at the lower level well into the depth of the deck, and at the upper level to keep out detritus and some water.

9.2.5.3 Installation of Nosings

The following notes may be helpful to the engineer proposing to design, specify or supervise nosings for new bridges or for the repair of old nosings (Fig. 9.11):

(a) For stability, the nosing should be at least 1·75 times, or preferably twice, as wide as it is deep. Where repairs are needed, the depth of the existing nosing should be ascertained from trial holes to enable the adjacent asphalt to be sawn on the correct line the first time.

(b) Where repairs to an old, failed nosing are required, the condition of the road asphalt should be considered carefully. If it is worn, cracked or tracked, consideration should be given to its renewal before the nosing is repaired. The adjacent asphalt is used to profile the surface of the new nosing, and if mis-shapen will give a distorted

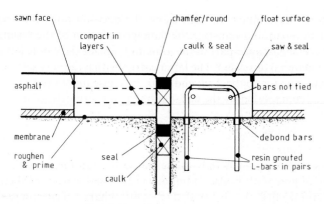

Fig. 9.11.   Nosing construction.

surface to the new joint. Further, renewal of asphalt up to a sound nosing is difficult and tedious, the scarifying and reinstatement being undertaken by hand. The use of machines for these operations would be very likely to damage or dislodge the joint.

(c)   A fundamental design decision is whether or not to include reinforcement to assist in anchoring the nosing to the deck. Many manufacturers consider this unnecessary and it is doubtful whether it can be justified in stress/strain terms. However, should a failure occur, especially on a high speed, heavily trafficked road, then the anchor bars do provide some level of restraint to blocks of nosing material which might otherwise jump out of the joint to become a hazard to traffic. If this insurance policy is chosen, the premium to be paid is a substantial increase in the work, the time taken and the cost.

(d)   In preparing the substrate, the system manufacturer will normally specify scabbling. This should, however, be lightly done—just enough to indent and clean the surface. Alternatively, if the manufacturer will agree, sand blasting may be preferred. Heavy scabbling to achieve a coarse-textured substrate is almost certain to destroy the integrity of the surface by loosening surface aggregate and cracking the matrix. A nosing cast upon such a shattered surface is likely to debond—not at the interface, but a few millimetres below it. The client will then have to pay for the subsequent repair because it was the concrete, not the nosing which failed. Proving that the contractor destroyed an otherwise sound substrate is difficult.

(e)   Where reinforcement is to be incorporated, stirrups should be in the form of pairs of L-bars rather than inverted U-shape bars. Many of the holes will not be completed where they were first intended to be, because the driller will encounter reinforcement in the deck concrete and have to move to another position. After drilling and blowing out the holes, the bars are grouted in with suitable resin. When cured, the bars should all be checked for anchorage bond. Many failures have been noted at this stage. If longitudinal 'distribution' steel is added through the stirrups, it should not be secured by wiring but laid in loosely as the placing of the mortar proceeds. If wired, it will vibrate during the mechanical compaction of the mortar and generate a hole in which to lie unbonded.

(f)   Shuttering must be securely fixed and remain firm. If a removable shutter is used, it must have a smooth face so that it can be removed without undue force. If an expendable shutter, e.g. of expanded polystyrene, is used it should be well anchored into the joint gap and supported to keep it straight and upright during the casting of the first nosing. The practice of casting both nosings together on each side of the polystyrene may be acceptable for short length repairs, it is fraught with risks for full-length nosing joints.

(g)   Priming of the substrate, adjacent asphalt and reinforcement should be thorough. Dusty concrete is difficult to wet with the treacly resin. This operation is aided in cold conditions by warming the substrate and resin immediately beforehand.

(h)   Materials for mixing will arrive on site pre-measured: the resin and hardener in tins, and the aggregate in bags. Because each batch for mixing is small, it is imperative that care is taken to get all of the contents of the tins and bags into the mixer. Especially in cold weather this is not easy because of the viscous consistency of the resin. Pre-heating the tins on a water bath will make handling easier.

(i)   Mixing of the small batches is best done in an open pan mixer where the process can be carefully monitored by eye. Scrapers should fit closely to the sides and bottom of the pan to ensure that all of the batch is included in the mix. The mixer should be located as near as possible to the joint to minimise the delay in getting the fresh mortar into the work.

(j)   The mortar should be placed carefully on to the freshly primed substrate and spread along the trench. The mix will be semi-dry, and resinous mixes will be sticky. As such, compaction is not possible with a poker vibrator, but is best performed using a small pneumatic hammer fitted with a foot plate. Compaction should be in layers not exceeding about 50 mm, with later layers following very quickly behind earlier ones.

(k)   When the trench is filled, the surface must be floated to a smooth, close-textured finish. Arrises will need to be chamfered or pencil-rounded. This can be achieved either at this stage or by using a portable grinding disc after the nosing has hardened and the shutter been removed.

(l)   Adequate time must be left for proper curing. This will vary widely from one to ten days, depending upon the material used. The manufacturer's requirements must be met with regard to both removal of the shutter and curing arrangements.

(m)   Surface preparation for poured mastic joint seals should preferably be by light sand-blasting, following which the joint should be caulked. The caulking must be deep enough to permit a full depth of sealant. Many past failures of sealants can be traced, at least in part, to skimped depth of material. The nosing faces are then ready for priming, which should take place while the sealant is being mixed.

(n)   After mixing, the sealant should be poured gently (to avoid air inclusions) into the gap against freshly primed faces. Self-levelling materials will flow, long after they are poured, down the crossfall of the deck to form a pool at the lower end and leave sub-standard depths elsewhere. Whilst the casting of nosings in warm weather is generally advantageous, the sealing of the joint in very hot conditions should be avoided. A seal poured whilst the gap is at its minimum will spend the rest of its working life in tension—the worst possible condition for it. There is some merit in high summer in leaving the joint sealing until the weather cools. Adequate time should be allowed for curing. Where repairs are being undertaken, sealing the joint is the last operation before re-opening the road to traffic. There are usually very strong political as well as commercial pressures to clear the site quickly, taking whatever short cuts are available, particularly cutting curing times. These must be resisted if a successful seal is to be achieved.

The foregoing notes refer generally to epoxy mortar nosings, although much of the contents can be applied to concrete and polyurethane. These are the materials currently available in the UK although elsewhere in the world, notably in France, Belgium and the US, a third generation of nosing material has been developed since the early 1970s. This is known as 'elastomeric concrete'[5] and comprises a stone aggregate and sand filler embedded in a vulcanised rubber matrix. The materials are delivered to site dry and pre-bagged, then mixed hot and vulcanised *in-situ* to form the nosings. Performance over more than a decade has been excellent. The shock-absorbing vulcanised rubber nosings are well bonded to their substrates and there is a much-reduced vibration transmission to the structure. In addition, a built-in rubber waterstop can be provided which is flexible and gives good waterproofing of the joint.

9.2.6 Prefabricated Joints

Where movements in excess of about 60 mm are expected, the use of some form of prefabricated joint is essential. These joints may also be chosen in preference to nosings for smaller movements where elastomeric concrete is not available.

Because the joints are made up of prefabricated units, they must be anchored in some way to the concrete deck or substructure. There are two different approaches to this anchorage problem: to cast anchorage ties into the base concrete at the time of construction, or to drill the hardened concrete and fix the joint units with anchorage bolts. The former is applicable essentially to new works whilst the latter may be used for either new or repair works.

In making the choice at the new design stage, the engineer must consider whether a repair or replacement is likely to be required during the working life of the bridge. If this is a reasonable probability, then to choose cast-in anchorages is not wise. Repair or replacement of such joints requires the breaking out of the anchorages, involving the breaking out and replacement of significant quantities of deck concrete. Such work under service conditions involves long periods of traffic lane possessions, with a prolongation of all the attendant disruption with its consequent direct and community costs. On the other hand, the design engineer in making the choice must consider that the most frequent reasons for the failure of prefabricated joints is failure of the anchorage bolts, beddings or transition strips (Fig. 9.12). These are all features of the bolt-down type of joint anchorage.

The choice of joint system should also have some regard (so far as this may be possible) to the question of obsolescence. The joint depicted in Fig. 9.13 was installed in the mid-1970s, and replaced the original which was of

Fig. 9.12.   Bolted-down prefabricated joint units. The original mechanical joints were replaced with these elastomeric units about ten years ago. Various repairs have been needed to the transition strips and adjacent asphalt. Some units are now loose.

Fig. 9.13. Prefabricated joint (now obsolete). The rubber seal is cracked and lacerated. The road has been surface-dressed to provide additional skid resistance, but the surfaces are not everywhere flush with the joint.

early 1960s vintage. By the late 1970s the present joint was no longer available, having been superseded on the market by an elastomeric unit system similar to that shown in Fig. 9.12.

Design of the built-in anchorage type should have regard to the difficulty of compacting concrete below flat metal bearing members where small air bubbles, excess water and laitance collect during concreting. Such nominally flat undersurfaces should be slightly sloped transversely to the joint, with corners rounded. An epoxy resin coating can be used to create slopes and at the same time give protection to steelwork. Anchorage ties should be fully welded to bearers, with welds located transversely to the line of the joint members to reduce the likelihood of fatigue fracture of the welds.

Bolted down joints rely heavily on three features for their success (Fig. 9.14):

(a)   the bolts, nuts and washers;
(b)   the bedding below the joint units;
(c)   the transition strip between the joint units and the road asphalt.

The second and third of these are needed to take up dimensional differences between factory-produced units made to mechanical engineering tolerances and the site construction of the civil engineer working to much larger dimensions.

Bedding materials are usually epoxy mortar or polymer modified concrete, the latter being much cheaper and generally preferable for depths in excess of about 10 mm. Preparation of the substrate and all subsequent processes are broadly similar to those for nosings.

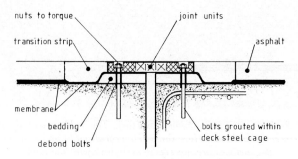

Fig. 9.14.   Bolted prefabricated joint.

9.2.6.1 Fixing Bolted Joints

After the bedding has hardened, the anchorage bolt holes are drilled. Positions have to be located accurately, either by using a template or by assembling and using the units themselves. Rotary percussion drills are preferred because the holes have slightly roughened internal faces. However, the driller must be prepared to change to a diamond bit when the inevitable encounter with the deck reinforcement occurs. He cannot change the position of the hole. The hole must be deep enough, and the bolt long enough, to enable the bolt to be debonded through the bedding and through at least the first 10 mm of concrete, which will offer little resistance to pull-out. The longest possible length of debond should be allowed to distribute the strain in the bolt. Holes must be clean and dry when the bolts are grouted. Anchorage bolts, nuts and washers should be of stainless steel.

After the bolts are securely grouted, the joints may be fixed, care being taken with whatever provision has been made by the manufacturer and designer for waterproofing. Few prefabricated joints are truly waterproof, but the efficiency of the provison made will depend largely upon the workmanship of joint bedding at this stage. Setting the joint in accordance with the temperature of the structure is at best an optimistic process, and unless the work has to be carried out in very hot or very cold conditions (both of which are undesirable for many reasons), it is better to use a bigger joint unit which has enough tolerance built in to render unnecessary the tedious intricacy of 'setting' the joint. After the nuts have been tightened to the specified torque, they should be checked and retightened, if necessary, after say 24 hours, in case the highly stressed resin grout in the bolt holes has relaxed.

It is normal, during both new works and repairs, for the fixing of a bolted joint to be carried out in a trench through the surfacing asphalt which will be used as a datum for the fixing of the units. This trench must be wide enough to allow generous working space around the units, and to permit stable proportions of about 2:1 for the transition strip. This will be filled with epoxy, polyurethane or cementitious mortar, and is yet another vulnerable part of the joint system. Forming this infill strip should be done with all the same care as for nosings and joint beddings. Furthermore, the strip will be located on top of the waterproofing membrane, so that it will not be bonded directly to the deck or joint bedding. A common tendency is for the asphalt to pull away from the transition strip during service, leaving a gap for detritus to collect. This may be countered by sawing a groove along the junction and filling it with mastic sealant.

It should be noted that few prefabricated joints are free from maintenance problems, and they should be inspected frequently. For this reason, some of the older joints with cover plates designed to 'keep out the dirt' need particular care and attention because they are so difficult to inspect.

9.2.7 Repair of Joints—Contractual Matters

For the engineer preparing to carry out the repair or renewal of joints of various types, particularly on busy arterial roads, the following general notes may be of value:

(a) A considerable proportion of maintenance work has to be effected over weekends, so that careful forethought and good preparation on the part of both the engineer and the contractor are important. Much of the work is weather-sensitive, and in rapidly changing climatic conditions the planned week-end work may need to be called off at the last moment. It is good practice to have available a contractual rate for frustration after the contractor has mobilised his labour and plant and cannot use it.

(b) Similarly, for weekend or overnight work, it is worth specifying that the contractor has available, at standby rates, items of key plant which, should there be a breakdown, could not readily be mobilised out of normal working hours.

TABLE 9.1
Time Allowance for Joints Repairs

Type of joint		Time to repair each lane or lanes in one possession period
Buried joint		Up to 2 days, depending upon amount of surface preparation
Bridging joint		12 hours
Nosings: epoxy		7–10 days ⎫
polyurethane		2–3 days ⎬ depending upon, *inter alia*, whether
HAC concrete		2–3 days ⎪ reinforcement is included
OPC concrete		10–14 days ⎭
Prefabricated joints:	Bolted	10–14 days
	Built-in	up to 30 days

(c) The length of time to effect a typical repair varies widely according to the joint type and materials used. Table 9.1 gives some general guidance on this matter.

(d) The system designer or manufacturer will have invested very much time, effort and money in the research and development of his product. He owes it to himself to ensure as far as possible that his product is installed and used as he intended. He can do this only if he has an active contractual presence on the site while his joint is being installed, acting as an aide to the resident engineer. This site presence will cost him money which, in the UK 'cheapest cost' tendering processes, he will not be able to afford unless his competitors include this cost also. The only possible way to achieve this end is for the engineer to specify this presence on site in the tender document. To do so is to have the best possible assurance of technical success with highly specialised materials and systems.

9.3 SUBSTRUCTURES

9.3.1 Reinforced Earth

Experience and developments in the practice of reinforced earth, originally used for retaining walls and earth dams, has enabled this technique to be utilised for comparatively cheap bridge abutments founded on poor ground which previously would have been extremely expensive to deal with.[6] Here again, advances in deck analysis methods provide the ability to predict reliably the effects on the decks of relative twist caused by differential settlements of abutments.

It must be understood that the term 'Reinforced Earth' is a trademark of Henri Vidal and the Reinforced Earth companies. However, within the context of this chapter, the term 'reinforced earth' is used loosely to mean the practice of incorporating horizontal layers of tensile material within the body of earth retained, so as to increase its strength and provide an anchor or tie for the facing material retaining it.

The early pioneering work of Vidal in this field was based upon the concept of layers of earth each supported at the face by horizontal strips of sheet metal (Fig. 9.15). The facings were held in place by flat metal strips which extended horizontally into the retained earth, and which were held there by the frictional component of the weight of the earth above them. The facings could accommodate settlement and compaction movements by their own deflections. Later developments have included more ornamental

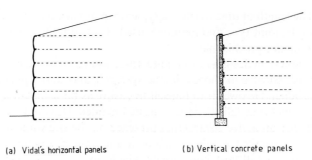

(a) Vidal's horizontal panels   (b) Vertical concrete panels

Fig. 9.15.   Reinforced earth wall sections.

facings of moulded glass reinforced plastic and more robust concrete panels arranged as small interlocking units, as vertical ribbed slabs (Fig. 9.16) or as flat horizontal panels lying between steel soldier supports which are tied back into the block of earth (Fig. 9.17). It is, of course, imperative when using rigid panels of any shape to make provision for subsequent earth movements, whether caused by compaction of the earth being retained or by ground settlements or subsidence. For example, the

Fig. 9.16.   Reinforced soil retaining wall. The ribbed concrete vertical panels are secured to the earth reinforcement with sliding toggles to permit settlement of the retained embankment.

*Fig. 9.17.   Reinforced soil wall under construction. The far wall and embankment are
partly filled. Steel soldiers are being erected in preparation for the placing of concrete
wall panels which will be slid into the recesses of the steel sections.*

anchorage of the earth reinforcements or ties to the facings may be in the
form of vertically sliding toggles. In addition, joints between adjacent
panels must be caulked with some suitable compressible material, so that
the concrete cannot bind, spall and crush.

There is perhaps a need to distinguish here between reinforced earth
walls and anchored walls.[7] The former utilises the weight of retained earth
to generate in the horizontal layers of reinforcement frictional resistance to
the forces trying to push over the facing material. Anchored earth
techniques rely upon horizontal bearing pressures developed at individual
anchorages located deep into the retained material. Both techniques are
suitable, and have been used for bridge substructures.

9.3.2 Mining Subsidence

In areas where extensive mineral extractions are likely to generate
significant ground movements, it has been the practice to design structures
to articulate freely, with provision in the design of the substructures for
jacks to be inserted to relieve twisted decks and to re-establish road profiles
after subsidence is complete. These have generally proved successful.

However, the surface areas of influence of deep mining of thick coal have been found to be greater than predicted.

It is worth noting that sometimes the forecasts by the National Coal Board of their dates and directions of mining have proved subsequently to be incorrect. It must be recommended therefore that the worst possible direction of mining should be assumed for a particular bridge, together with the worst possible timing.

As with settlement of supports founded on poor ground, it is preferable, if it is feasible to do so, to design the deck of the structure without joints. In such cases, reinforced earth should be considered for abutments and wing walls.

9.3.3 Footbridges and Subways

Particularly in urban areas, traffic policies which require the segregation of pedestrians from major flows of vehicular traffic pose the frequently difficult problem of how to arrange for pedestrians to cross the road safely. There is in this matter a fundamental issue of whether the pedestrian or the motorist should be diverted to facilitate the passage of the other. Setting this aside, it has generally been the accepted practice, based on financial considerations only, for vehicles to take precedence over pedestrians. Having decided this, then there are two possible solutions to the problem: to lead the pedestrians either overhead by footbridge or below ground by subway. Both of these options require structures which have developed in the light of experience.

Whilst many early footbridges were purpose-designed for their individual sites, the more recent trends have been towards standardised designs adapted for specific sites. The reasons include speed of design and detailing, and, in the UK, speed also of technical approval by the Department of Transport. The Department's policy has been to encourage standardisation, and to discourage individual designs, by making technical approval a complex process. The adaptation of standard footbridges to individual sites has its greatest effect upon the substructures. These need to provide the shortest, easiest possible route for the pedestrians. Long, steep or folded ramps lead to user resistance, with pedestrians preferring to risk life and limb dodging traffic rather than to undertake an exhausting climb (Fig. 9.18). Pedestrian facilities can be improved if vehicle underpasses are only partly depressed, with the sides raised artificially with bunds. Jones[6] suggests that the cost of this option is of the order of two-thirds that of a full cutting, and less than half that of a viaduct. However, the cost may still be some 2·5 to 3 times that of an 'at grade' highway.

Fig. 9.18. Pedestrians and traffic. Some pedestrians prefer to climb over the pedestrian barrier and dodge the traffic rather than climb the lengthy ramps to the safety of the footbridge.

Rapidly escalating social mis-use of pedestrian subways, including crimes of assault in various forms, has necessitated changes in this form of pedestrian facility to make them socially more acceptable. At only a few short, well-used crossings is the formerly simple rectangular box now an acceptable form (Fig. 9.19). There is a social need to use a wider span, which gives pedestrians a feeling of openness and of greater security. This may be achieved by using abutments flared out in plan, or by increasing the span and using bankseats in lieu of abutments (Fig. 9.20).

Bold planning developments in some large city centres have included the provision of shops and kiosks in subways. Not only does this make use of valuable land space but also it helps to make the subways safer, more pleasant places for people. There has, however, been only limited success in these ventures. Traders are not keen to work in unpleasant subterranean environments. Furthermore, the technical inability of the civil engineer to provide adequate waterproofing, especially joint sealing, has led to much friction following ruined stock, and/or limitations on the available useful space. In consequence, the retail outlets provided have not always been taken up.

*Fig. 9.19.   Early pedestrian subway. Because of its length, this mid-1960s subway
was built 600 mm wider than the standards then in force. It has been converted recently
to provide a cycleway.*

*Fig. 9.20.   A pedestrian subway becomes a small bridge. Subway of the mid-1980s,
having wide walkway, sloping brick-faced abutments, lights beyond the reasonable
reach of vandals.*

9.3.4 Abutments

The practice of creating rectangular spans with vertical abutments is still valid for many structures. However, there are some sites at which the best possible sight lines are needed for traffic passing through the bridge. To provide these without an unreasonable increase in span, the inclined abutment has been developed (Fig. 9.21). Although formwork is more difficult and expensive than for a vertical wall, the reinforcement is simpler because the wall can be mainly of mass concrete, with a substantial reduction in the back face steel required.

In detailing cantilever or inclined abutments, it is usual to provide 'distribution' or crack control reinforcement in the front (compression) face. In most bridges this is sound practice. However, where the front face is subject to much salt spray raised by fast moving traffic, and particularly where this face is sheltered from rain by the bridge deck, consideration should be given to the vulnerability of this reinforcement to penetration of the concrete cover by chlorides. Options include the sealing or coating of the concrete face with a waterproofing membrane, and the use of

Fig. 9.21. Inclined abutment. This particular span is wider than necessary. Full advantage of the potential sight-line is not achieved because the toe of the approach embankment spreads too far, amenity planting is too near to the abutment, and the horticultural maintenance stops short of the ideal standards for which the abutment was designed.

polypropylene grids to control cracking in lieu of steel reinforcement. No experience of the performance of these grids is available in this context, although they have been incorporated successfully into extra cover added during concrete repairs, where the original cover was found to be substandard (Fig. 9.22).[8]

Changes in techniques of handling and placing of concrete, particularly the rapid development of centrifugal concrete pumps, with their high lifting capability and long, flexible delivery hoses, have made possible the efficient placing of concrete in substructures in full height lifts, rather than in a series of smaller ones. This has, in turn, enabled designers to provide exposed concrete faces without horizontal rebates to disguise lift marks. Instead, formwork designed with upright striations, ribs or rebates, which disguise more easily the vertical stains and streaks caused by natural weathering, is now a practical and better alternative. It is, however, still the policy of many designers of bridges in environmentally sensitive locations to use brickwork, masonry or blockwork facings applied over rough shuttered concrete, reinforced earth or close-piled walls. This facework is now either free-standing or tied with wall ties across an open drainage gap. Earlier

Fig. 9.22.   Polypropylene grid reinforcement for cover concrete. The plate shows the repaired concrete face being saturated prior to the spraying of an extra thickness of cover concrete. The grid is fixed with nylon spacer studs.

Fig. 9.23. Cracked facing brickwork, showing a horizontal crack in the joint of integral facing brickwork at the level of concrete backing lift.

walls of integral facings, or with gaps filled with mortar, caused problems of shrinkage cracking reflected in the horizontal joints (Fig. 9.23), or extensive unsightly lime staining of the facings.

The decorative painting of concrete substructures can give excellent appearances if kept well maintained. All too often, however, they are left neglected and become unsightly (Fig. 9.24).

9.3.5 The Bearing Shelf

In the design of new bridges, thought and care needs to be given to the question of accessibility for inspection and maintenance of bearings, and this should be addressed to the bearing shelf as a whole. At free ends, especially where movements are substantial and hence the bearings are deep, this normally is not difficult to achieve. A clearance between deck soffit and shelf of 100 mm should be considered an absolute minimum, with 150 mm providing a more reasonable working space. At fixed ends this will not normally be achieved unless positive steps are taken (Fig. 9.25). Consideration might be given to the inclusion of a downstand below the soffit of an *in-situ* deck, or to the use of bearings which are unnecessarily deep for the movements to be accommodated. However, it must not be

Fig. 9.24. Neglected decorative painting of viaduct piers. Originally coloured black, contrasting dramatically with the white exposed aggregate fascias, these viaduct piers now have an untidy, neglected appearance.

forgotten that the lower the fixed end fixing, the greater the horizontal dynamic movement in the joint gap at the road surface.

The design of the bearing shelf should include provision for surface water from the deck, which is likely sooner or later to penetrate the movement joints above. For some 15–20 years, bridge designers have been incorporating secondary drainage channels at the backs of the bearing shelves. Frequently these have been too small and easily become blocked with detritus. There is a growing body of opinion that the better way is to drain towards the front face (Fig. 9.26), where any collecting channel can be kept clean more easily, and where degradation from penetration of the cover concrete is less likely, less damaging and more easily repaired. There is indeed a good case for having no drainage channel, letting any percolation water spill over the structural face. Because the abutment can be slimmer, this provides a slightly cheaper solution where the resulting stained face does not matter.

The incorporation of some form of flashing or drip mould is desirable to prevent percolation and condensation water from travelling along the deck soffit to the upper level of the bearings.

Fig. 9.25. Inadequate space for bearing shelf maintenance. Typical of fixed end supports, it is not possible to maintain a bearing shelf having only 12 mm of clearance.

9.3.6 Bearings

Some twenty or so years ago, it was popular for bridge designers to conceal bearings from view as much as possible, both for aesthetic reasons and to keep out dirt and nesting birds. Experience has shown that regular inspection and sometimes maintenance are necessary, so that bearings should be fully accessible for both purposes. Fashions have changed, so that bearings which look like bearings are now aesthetically acceptable. Experience also demonstrates that however well bearings are concealed,

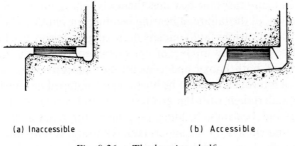

(a) Inaccessible   (b) Accessible

Fig. 9.26. The bearing shelf.

they are still subject to accumulations of detritus which, because of the concealment, lies undetected until something goes wrong. It is then difficult to remove the dirt from around the 'built-in' bearings.

Whereas bearings on abutments and piers can be exposed to view, those within halved joints in slabs cannot. This factor militates against suspended span forms of structure in favour of continuous decks.

The design and manufacture of bearings is now a specialist task undertaken by various manufacturers, depending upon the type of bearing used. This desirable specialisation has replaced earlier individual designs which frequently were prepared by inexperienced general bridge engineers, who did not always recognise the problems of bridge maintenance. In particular, the use of unprotected mild steel plates which can corrode and seize rapidly, and of thin bonded PTFE which can ruckle or wear quickly, are to be avoided. In one case the PTFE was so thin that, with only a little wear and with the rotation of the deck, the front edge of the upper and lower plates of the bearing seized in one direction of movement only. These plates consequently travelled horizontally relative to each other. This progressive movement, which occurred during a period of less than a year, would have ended with serious consequences to the bridge had the defect not been discovered during a routine bridge inspection.

The practice of having sliding or roller bearings made and delivered with a calculated pre-set has proved to be an over-theoretical complication during construction. It is almost impossible to predict at the time of ordering what the real conditions will be at the time of fixing the bearings. Neither can it be foreseen whether subsequent operations will render invalid the calculated pre-set of bearings already built into the structure. It is much better to use oversize bearings which have enough inbuilt tolerance to avoid the need for pre-setting.

9.3.7 Bearing Beddings

Although in many cases the bearings themselves have not failed, there are many instances of the failure of bearing beddings or plinths. These failures have arisen from various faults, mostly occurring in the construction stages. They include the faulty mixing of epoxy materials and the inadequate compaction of both epoxy and cementitious mortars. The bedding of bearings is a job which tends to be hurried and skimped on site, but which should be undertaken with the greatest care and conscientiousness. The removal of old bearings, bedding or plinths for renewal is a difficult operation, and should not be undertaken without a careful assessment of loads and deflections which can be tolerated by the deck without inducing unacceptable overstresses. Jacking arrangements, loads and movements must be specified in detail, and monitored carefully on the site before, during and after the removal and replacement work.

Where beddings and/or plinths have failed, removal of the old material together with the cleaning of the remaining surfaces may be carried out effectively and at once, utilising high pressure water jetting. This technique has been developed rapidly in recent years, and extremely efficient cleaning and cutting can now be achieved with or without the introduction of grit or slag to the jet stream. It is a highly specialised process, and the inexperienced engineer should carefully avoid any attempts to specify water pressures, water volumes or jet sizes, all of which are interdependent. The specialist operator will recommend these, depending upon the task to be undertaken.

To effect the bedding or plinth renewal, there are available various formulations of grouts and mortars. Wherever possible, cementitious materials which are tolerant of damp surfaces should be chosen. Many are designed to expand slightly—just enough to overcome the natural shrinkage of the material. However, expanding agents based upon powdered iron or aluminium should be avoided. In the former case the expansion is not readily controlled, and in the latter case the hardened material is likely to be slightly porous and therefore vulnerable to frost degradation. Although rebedding by hand compaction of semi-dry mortar is a valid method, it is very labour-intensive and difficult to supervise. For these reasons it is best suited to small numbers of comparatively deep sections. For thin sections, and all cases where many beddings are involved, prepacked self-levelling grouts, with their high flowability, are much more suitable.

9.3.8 Degradation of Substructure Concrete

Perhaps the most significant aspect of substructure performance during the last decade has been the behaviour of the concrete itself, both as a heterogenous material and as a passive bed encasing the reinforcement.

9.3.8.1 Alkali–Silica Reaction

Until the late 1970s, the possibility of UK concrete suffering from alkali–silica reaction (ASR) was considered unlikely, and no precautions were therefore taken. Indeed few UK engineers had ever heard of the problem, although it has exercised the minds of researchers and engineers in other places, notably the USA, Denmark and South Africa, for some 30–40 years. By the time the problem of, *inter alia*, Marsh Mills Viaduct in

Devon received extensive media coverage,[9] the deck of a major footbridge in Birmingham, West Midlands, had already been replaced, and the afflicted precast prestressed beams had been dispatched to the Cement and Concrete Association and the Building Research Station to further their investigations.

Just as corrosion of steel requires both moisture and air to progress, so does ASR. There is no practical way in which these elements can be kept away from bridge structures, so that once the process is commenced, it cannot be contained. In the UK the Concrete Society produced its guidelines[10] in 1985 on the avoidance of ASR in future works. Whilst clearly it is possible to choose aggregates with care to avoid reactive contents, it is nevertheless difficult and uncertain. The better way appears to be the control of the alkalinity of the cement.

It should be noted that the presence of ASR in the concrete of existing structures is not of itself cause for despair. It is possible for the active ingredients in the concrete to become exhausted before damaging expansion can occur. In such cases, the small amounts of gel produced are contained within the interstices of the aggregate and matrix. In cases of doubt, long cores should be taken out and immersed in warm water in the laboratory. This will encourage the ASR to proceed apace if reactivity is at all possible. The lengths of the cores should be monitored for expansion over several months, and the extent, if any, of latent ASR can thus be assessed. If there is much activity still to develop, some time will have been bought to consider the implications for the structure. Otherwise no further action need be taken.

There does appear to be some connection between the presence of ASR and chloride contamination. The reasons are not currently understood, although it seems possible that because both salt and the ASR gel are hygroscopic, they aid one another by encouraging the structure to hold more moisture than it otherwise would.

9.3.8.2 Frost

Some of the damage to substructure concrete experienced in recent years is not attributable to either of these causes but to frost action, particularly in the more exposed locations such as the ends of pier crossheads. These become saturated, and never dry out because the salt (and perhaps ASR) holds excessive moisture in the concrete.

Unlike their road engineering colleagues, bridge engineers generally have been reluctant to take advantage of the frost resistance offered by air-entrained concrete for structural work. It should be noted, however, that the incorporation of an air-entraining agent in the mix at an appropriate rate of, say, 4·5% (± 1%) will require a reappraisal of the mix design with, in particular, a substantially reduced fines content. Additional vibration of the concrete during placing will also be needed to work out the voids, particularly at the shuttered faces. Experience of obtaining from a pre-mix plant a guaranteed admixture dosage in the modified mix has shown that there may be a significant financial penalty involved. The engineer must consider whether freedom from the risk of frost damage to the bridge is worth the extra cost.

9.3.8.3 Sulphates

Careful examination and testing of decayed concrete in several early 1970s bridges at Peterborough, UK,[11] revealed that excessive sulphates from road de-icing salt were at least partly responsible for the degradation. The mechanism in this case is not the well known rusting of the steel reinforcement associated with chlorides, but an attack on the chemical structure of the matrix. The result is a severely weakened concrete.

9.3.8.4 Chloride Attack

The most widespread cause of reinforced concrete degradation in bridge substructures is the de-passivation and consequent expansive rusting of the steel reinforcement. This action is caused by the acidic properties of the choride ions from the soluble sodium chloride content of road de-icing salt percolating through movement joints, and from high-speed vehicle spray. Salt in the atmosphere from marine sources is of comparatively little significance in this context. The major damage locations are bearing shelves and pier heads of simply supported bridges and viaducts.

9.3.8.5 Distribution of Degradation

Many engineers have taken great care with carriageway movement joints, and this is often apparent in the performance of the substructures. Equal care and emphasis is not usually given to footways, verges, services troughs and central reservations. In consequence, it is normal to find that pier heads are saturated and degraded extensively under the sides and centre of the structure, and much less affected under the carriageways (Fig. 9.27). In the past, design engineers working under pressure have studied the joint manufacturer's literature. This always describes and illustrates clearly what is recommended for carriageways, but normally deals with the rest of the structure in cursory fashion. Little thought is therefore given to what is the most difficult part of the movement joint design so far as sealing is

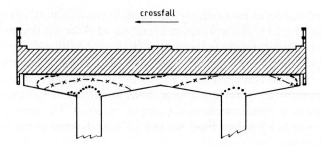

Fig. 9.27. Distribution of chloride attack and effect of repairs. −350 mV contours: ⋯⋯, before repairs; −×−×−, after repairs; −−−−, 1 year after repairs.

concerned, and much is left to chance. Services troughs are particularly difficult because they tend to collect water from a wide area and conduct it towards the joints.

Whilst the major source of salt penetration damage to substructures is from water percolating through movement joints, other damage occurs to underpass retaining walls, bridge abutments and piers from salt spray raised by fast moving traffic. The worst affected areas are up to about 2 m above the kerb or road level, particularly on bridge abutments sheltered from rain by their decks. In underpasses where adjacent retaining walls are subject to similar spray conditions (Fig. 9.28), but where surfaces are frequently washed by winter rain and snow, the chlorides have little chance to become concentrated or to penetrate the cover concrete, and comparatively little degradation occurs.

9.3.9 Comments on Concrete Repair Processes

The subject of repairs to concrete degradation is dealt with in Chapter 4, and only a few comments appear here. It cannot be emphasised too strongly or too often that in this activity good conscientious workmanship is of the greatest importance. The world's best repair materials, carelessly applied to surfaces poorly prepared, will be wasted. It is important too in this, as in other highly specialised fields of bridge engineering, that the repair system supplier be implicated contractually on the site by being represented technically while the system is being applied. The representative must agree formally such matters as the adequacy of preparation— how clean is 'clean', how dry is 'dry', how rough is 'rough'—methods of mixing and application of the materials, and the ambient conditions in which the work is being carried out.

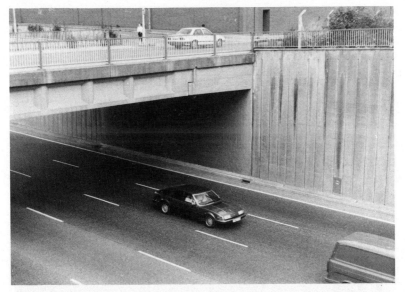

Fig. 9.28.   Abutment face subject to salt spray. The whole width of the sheltered abutment face became degraded by salt spray, whereas adjacent exposed wall faces were little affected.

9.3.9.1 The Effect of Repairs on Structural Performance

Where major repairs to degraded concrete in a bridge are required, the engineer must assess, as far as he can, what effect the removal of substantial quantities of surface concrete will have on the structural behaviour of the members concerned. In many cases a subjective judgement will suffice to determine that no precautions need to be taken. Sometimes this is not obvious, and the engineer will need to specify the action needed, such as jacking, propping or removal of the concrete in alternate panels. Little guidance is available on this aspect of structural repairs, and research is urgently needed to establish some guidelines.

Where reinforcement, which was presumably under stress, is completely exposed and cleaned in preparation for rebedding in cementitious mortar, it may be totally relieved of stress during this process (Fig. 9.29). One may assume that if the structure has been jacked to relieve it of dead load, or not propped but protected from receiving live load, then after completion of the repair it may be available to take these loads. In cases where props are not used, an alternate panel sequence of repairs may be specified (Fig. 9.30). Whether this genuinely helps is unknown, but it costs much more time and

Fig. 9.29. Exposed reinforcement prior to repair. The hook on the main bar is there because the designer was short of bond length! The stirrups also have been relieved of stress.

Fig. 9.30.   Alternate panel repairs. Panel 27 is being repaired except for a portion below the bearing which will follow later. Panels 26 and 28 will be repaired when 27 has cured.

money to take this precaution on the assumption that it does. Reliable information on such matters would be helpful in a field of activity where assumptions might prove hazardous.

9.3.9.2 Monitoring of Repairs

Before repairs are undertaken on a large scale, it is prudent to undertake a corrosion survey including, *inter alia*, half-cell potential mapping of the affected surfaces.[12] It is normal practice to take the half-cell readings at the nodes of a grid. If permanent markers are set up, the grid may be reproduced accurately after the repairs have been completed. The surfaces can then be monitored non-destructively to ascertain the effectiveness of what may be a comparatively temporary repair because of the difficulty of eradicating completely the cause of the degradation. It is assumed here that the objective of the repairs in corrosion terms may be considered to be to achieve and sustain reinforced concrete having electrode potential values higher (i.e. less negative) than $-200\,mV$, the level above which there is a 95% probability that corrosion of the reinforcement will not occur.[13]

Many maintenance engineers think that their repairs will last a lifetime, or are under great pressure to do work rather than monitor structures. Revenue money for maintenance work is always difficult to obtain, and costs must be kept to a minimum. However, the cost of monitoring is small, although the detailed analysis of results is expensive in time. Little work appears to have been published in this field, although some half-cell monitoring of viaduct pierhead crossbeams is in hand in the West Midlands, UK.

At one site, after repair of the movement joints, total crossbeam side faces, ends and soffits were stripped to expose the face reinforcement, and new concrete faces were put on. The bearing shelves, being inaccessible, were not repaired, but were thoroughly washed and flooded with potassium silicate solution to seal them. The monitor showed that shortly after repair, widespread values of half-cell readings were obtained. It took about a year for these to reduce to minimum values spread over a much narrower band of about $\pm 50\,mV$. Not all of this band lay above the $-200\,mV$ level, indicating that there is still potential for further corrosion in this beam.

At another site, partial repairs only to crossbeam faces were undertaken. The areas selected for repair were generally rectangular in shape, drawn up to ensure the removal and replacement of surface concrete where electrode potential values less (i.e. more negative) than $-200\,mV$ were recorded. The monitor at this site therefore covers both repaired and unrepaired concrete, providing some measure of control. Over a two-year period to date, some broad features have emerged concerning the reproducibility of results:

(a)   Readings may vary widely at any one time and location, although a variation of up to $\pm 20\,mV$ may be considered normal. Readings should therefore be taken two or three times to obtain a mean or 'characteristic' value.

(b)   Characteristic values so obtained in the morning may not be confirmed in the afternoon of the same day within a similar margin.

(c)   There is a broad trend for characteristic values for unrepaired concrete in late summer to be 'optimistic', and in mid-winter to be 'pessimistic', by up to about $\pm 30\,mV$. Examination of records from the local Meteorological Office provided some indication that this trend mirrors changes in average humidity during the periods immediately before and during the days of the readings (Fig. 9.31).

(d)   When this seasonal shift is applied to readings on the repaired concrete, it appears to give an over-correction, probably related to the differences in permeability between the repair mortar and the

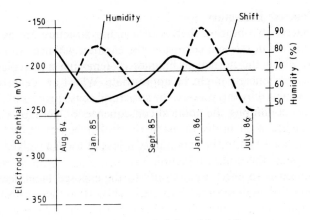

Fig. 9.31.   Seasonal shift and humidity.

parent concrete. However, the wide scatter of results shortly after repair again reduced after more than a year to a band of about ±40 mV, generally in this case above the −200 mV level (Fig. 9.32).

It must be emphasised that the foregoing does not represent better than an indication of broad patterns and trends at only two entirely separate structures, where different repair systems have been used. Much more work needs to be done before firm guidelines can be produced for the interpretation of half-cell survey data. It is sufficient here to warn that a single set of single readings may be misleading, and of little better value than a broad qualitative guide to identifying areas of worst corrosion.

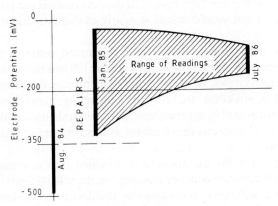

Fig. 9.32.   Progress of repassivation following repairs.

9.3.10 Concrete Protection

Where inaccessible bearing shelves of a major structure are extensively degraded by reinforcement corrosion, the engineer will need to assess the extent to which the normal techniques of removing and replacing the contaminated concrete might be appropriate. Whilst an idealistic repair might be achieved by this process, the structural consequences of changing the support conditions, the political consequences of, perhaps, closing or limiting the use of the bridge to enable the deck to be jacked, and the high costs both direct and to the community if loss of use is involved, may all militate against this course of action.

One option is to effect repairs only to the exposed faces, leaving the inaccessible shelves unrepaired. Experience shows that the expansive forces of corrosion cannot spall the concrete upwards below the bearings. The shelf between bearings is structurally unimportant. Although this is a lowest cost option, the corrosion cells will remain active, so that further face repairs will be required from time to time. It may, however, be possible to slow down the action by sealing the surface against further ingress of water and contaminants.

Alternatively, but at much greater initial expense, following minimal surface concrete repairs, it may be worthwhile to attempt to stabilise the corrosion of the steel by using a system of cathodic protection. In this case the repair system chosen should match as nearly as is practicable the parent concrete in the structure. This is particularly important where part faces have to be patch-repaired, because the effectiveness of cathodic protection depends upon the even or uniform distribution of the impressed current over the whole structural face. The resistance to this current is controlled significantly by the moisture content of the concrete. Hence the porosity of the repair material should match as nearly as possible that of the parent concrete.

Cathodic protection techniques for reinforced concrete bridges and viaducts have been developing, notably in the US and more recently in the UK. Various systems of distributing the impressed current over the surfaces of a structure are available, including a network of cables subsequently covered by sprayed cementitious mortar, and conductive paint films. At this stage in this development, it is premature to comment upon performance.

Concrete cover for bridgeworks is specified by the engineer on the guidance of the current codes of practice—in the UK BS 5400.[14] The cover requirements will vary according to the location, environment and sensitivity of the structural element considered. However, experience has shown that after only 15 years of service, significant levels of chlorides may be present up to 150 mm below the surface in the highly aggressive micro-environments of abutments and pier heads below leaking movement joints. Although it is difficult to prove it, this is almost certainly true also of half joints in decks. It is therefore imperative that in such locations concrete cover as we normally understand it is not considered to be a protection for the embedded steel, but solely a structural embedment to generate its interaction with the concrete. To achieve protection, the cover needs to be supplemented with some form of coating or sealant to waterproof its surface. This should be applied ideally to new structures in anticipation of joint leakage, and to repaired structures where further leakage may later regenerate corrosion. Whilst coatings may perform adequately and may be used, where access is good, for re-coating as an ongoing maintenance task, sealants which penetrate and chemically block the pores and microcracks of the surface concrete appear to offer a more reliable choice. Work carried out in the USA[15] identifies certain silanes, epoxies and methacrylates as providing potentially the best protection.

9.4 INSPECTION AND MAINTENANCE

9.4.1 Access

The need to be able to expose the bearings and shelf for inspection purposes has already been referred to in Section 9.3.5. Of similar importance is the need to be able to get easily to the bearing shelf to see on to it. It has been a popular practice in the past to pave the side slopes under bridges, in some cases taking the pavings right up to the bearing shelf. A much better practice is to build into the paved slope a flight of steps leading to an access walkway along the front face of the abutment about 1·2–1·5 m below the bearing shelf (Fig. 9.33). In addition, or alternatively, access steps down to the walkway from the upper level may be provided. River and canal bridges may not always enjoy the same flexibility, but consideration should always be given to the provision of a berm in front of the abutment from which to inspect the bearings. Alternatively, on larger bridges, this might take the form of an access gallery behind the bearings (Fig. 9.34), accessible either from the abutment ends or by manhole from the upper verge or footway.

Only rarely is there an opportunity to incorporate a permanent access walkway along pier heads, and more normally it is necessary to use hydraulic platforms or scaffolds for access. It is always appropriate to consider at the design stage the provision of space and hardstanding for such apparatus.

Fig. 9.33. Access steps to abutment. Originally built with no access provision, inspection and maintenance of the abutment and bearings was difficult. The steps and access berm have therefore been built into the paved bankslope.

9.4.2 Regular Inspections and Cyclic Maintenance

Whilst the need for regular bridge inspections has always been understood and generally practised, the process was nationally formalised in the UK by the publication of Department of Transport Memorandum BE 4/77.[16] Bridge designers have for long been urged to 'design with maintenance in mind', without having a very clear understanding of what that meant. Indeed, the introduction of Principal Inspections as set out in BE 4/77, and their subsequent results, crystallised in bridge engineers' minds the significance of regular maintenance, and the need to keep detailed records with photographs. Only by so doing can the subjective descriptions 'good condition' or 'fair' be evaluated, and deterioration monitored.

In particular, designers had secondary drainage channels placed on bearing shelves, with outlets or downpipes to drain water away. Some also built in good access arrangements. Few, however, set up proper management systems, including manpower and budgeting provisions, to enable and ensure the assumed and necessary cyclic maintenance took place. As a result, bearings stood rusting in pools of water, while drainage channels were clogged with detritus, and pigeons nested snugly in the warm polystyrene granules. In many cases there had been no maintenance and no

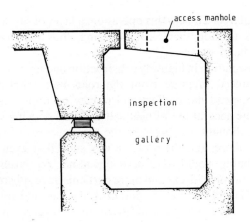

Fig. 9.34.   Abutment inspection gallery.

detailed inspection between construction and the first Principal Inspection some 10–15 years later.

In earlier eras, masonry arches required little maintenance, and all problems were readily visible. Most modern retangular spans have moving joints and bearings which, like all other moving machinery, need regular servicing if they are to be kept in good working order. It is interesting to speculate upon the extent to which chloride degradation of pier tops might have been delayed—or even possibly averted—by regular twice-per-year washing of the bearing shelves. It is now clear that it is good practice to do so. Similarly, the frequent hosing down of sheltered abutments and piers during the winter would avoid extensive chloride degradation from vehicle spray.

9.5   MATERIALS FOR DE-ICING

9.5.1 Rock Salt

Much of our concrete bridge maintenance of the 1980s stems from the use of rock salt for winter road de-icing. It is perhaps worth reviewing briefly the history of this use, and considering the alternatives.

Until the late 1950s, dealing with modest quantities of snow and ice was a normal winter driving hazard for which most prudent drivers prepared themselves. Highway authorities assisted by providing, and sometimes spreading, coarse clinker grit, which in those days was cheaply available. In rural areas particularly, stockpiles or bins were kept at the roadside, mainly at hazard spots, for drivers and highway lengthsmen to spread as conditions required. To help this operation, a layer of salt was spread over the stockpiles to stop them from freezing, thus making them readily accessible by hand shovel.

The rapid escalation in highway construction and use from the late 1950s onwards produced pressure from the road user lobbies for highway authorities to spend more effort and money on improving winter road conditions. The general use of neat salt for road de-icing was one result.

No serious thought was given to the consequences of this change for bridge maintenance, and the pressure was such that even if it had, the outcome probably would have been the same. The ensuing ravages of chloride penetration of decks and more particularly of substructures is now well known. The cost of making good this damage and trying to prevent recurrences is so vast as to make investigation into alternative de-icers very urgent.

9.5.2 Urea

Airport runways, where corrosion of aircraft by salt is even more serious in its consequences than that of bridges, have for long been de-iced with urea. Full scale trials in the UK in 1984/85 and 1985/86 on Midlands motorways having extensive viaducts were successful in de-icing the roads.[17] It is not yet known whether any damaging effects on the structures have been caused by the bio-degradation of the spent urea in cracks and crevices of the structures.

Anyone contemplating the use of urea for this purpose should study carefully the handling and storage problems of the urea prills, and the conversions which may be required to the spreading vehicles. They should also be aware of the additional ventilation required in subway and underpass drainage and pump chambers to disperse the ammonia fumes which emanate from the run-off water.

9.5.3 Alternative Materials

Despite extensive research, particularly that carried out in the USA, no reliably safe practical alternative to salt or urea has been discovered. There is an urgent need for one to be developed.

Whereas it may be possible to reduce or even eliminate the degradation of reinforced concrete in new structures by using protective sealants or coatings, the prospects of arresting decay in existing bridges does not appear to be good, while year after year we continue to subject them to aggressive de-icing materials. This is undoubtedly the most serious deficiency in concrete bridge performance over the last decade or so.

REFERENCES

1. Price, A. R., *The performance in service of bridge deck expansion joints*, Dept of the Environment, Dept of Transport, TRRL Report LR 1104, Transport and Road Research Laboratory, Crowthorne (1984).
2. Anon., Road holding grid cuts down repairs, *New Civil Engineer*, London (3 July 1986).
3. Brown, S. F. and Broderick B. V., *Polymer grid reinforcement of asphalt*, Association of Asphalt Paving Technologists, San Antonio, Texas (Feb. 1985).
4. Price, A. R., *The performance of nosing type bridge deck expansion joints*, Dept of the Environment, Dept of Transport, TRRL Report LR 1071, Transport and Road Research Laboratory, Crowthorne (1983).
5. Busch, G. A., *Experience with elastomeric concrete expansion joint transition dams in bridges, joint sealing and bearing systems for concrete structures, Vol. 2*, Publication SP-70, American Concrete Institute, Detroit (1982).
6. Jones, C. J. F. P., *Earth Reinforcement and Soil Structures*, Butterworth's, London (1984).
7. Jones, C. J. F. P., Murray, R. T., Temporal, J. and Mair, R. J., First application of anchored earth, *Proc. XIth International Conference on Soil Mechanics and Foundation Engineering*, San Francisco (1985).
8. Anon., Ring road trials may herald £4 m repairs, *New Civil Engineer*, London (21 Feb. 1985).
9. Anon., Concrete problems burst forth, *New Civil Engineer*, London (24 April 1980).
10. The Concrete Society, Minimising the risk of alkali–silica reaction, guidance notes and model specification clauses, Report of a Working Party; Alkali–silica reaction: new structures—specifying the answer, existing structures, diagrams and assessment, Papers for a *One-Day Conference*, London (11 Nov. 1985); draft Concrete Society Technical Report.
11. Anon., Symptoms of sulphate, *New Civil Engineer*, London (10 July 1986).
12. Vassie, P. R., *A survey of site tests for the assessment of corrosion in reinforced concrete*, Dept of the Environment, Dept of Transport, TRRL Report LR 953, Transport and Road Research Laboratory, Crowthorne (1980).
13. Van Deveer, J. R., Techniques for evaluating reinforced concrete in bridge decks, *ACI J.*, Detroit, **72**(12) (1975).
14. British Standards Institution, BS 5400, *Steel, concrete and composite bridges, Part 4: Code of Practice for the design of concrete bridges*, London (1984).
15. Pfeifer, D. W. and Scali, M. J., *Concrete sealers for protection of bridge structures*, National Co-operative Highway Research Program Report 244, Transportation Research Board, National Research Council, Washington, DC (1981).
16. Department of Transport, Technical Memorandum (Bridges) BE 4/77, *The inspection of highway structures*, Dept of Transport Highways Directorate, London (1977).
17. Anon, Midland Links urea trials to continue, *New Civil Engineer* (14 Aug. 1986).

Index